Pittsburgh Antique Radio Society Publications
David W. Kraeuter, General Editor

1. *A Bibliography of Frank Conrad,* (Third Edition), 2007.
2. *The U.S. Patents of Reginald A. Fessenden,* (Second Edition), 2007.
3. *A New Bibliography of Reginald A. Fessenden,* (Second Edition), 2007.
4. *The U.S. Patents of John H. Hammond, Jr.,* (Second Edition), 2007.
5. *An Interview with Harold Beverage,* Richard Brewster, (Second Edition), 2007.
6. *Radio and Television Reminiscences: Raymond M. Bell in the Pittsburgh Oscillator,* (Third Edition), 2007.
7. *Electronic Essays*, David W. Kraeuter, (Sixth Edition), 2012.
8. *The U.S. Patents of Harold S. Black, Jack S. Kilby and Robert N. Noyce*, (Second Edition), 2007.
9. *The U.S. Patents of Stuart W. Seeley* (with a bibliography of Seeley's writings), (Second Edition), 2007.
10. *Ten Patents from Radio History*, David W. Kraeuter, 2007.
11. *Frank Conrad's Radio Patents: The Complete Texts*, (Second Edition), 2007.
12. *Electronic Reviews: Hundreds of Thoughts on 100 Books,* David W. Kraeuter, 2008.
13. *The 3 Strikes Camp Stories,* Karl Laurin, 2008.
14. *Vintage Radio Redux,* (Second Edition), Karl Laurin, 2012.
15. *A Radio Patent Chronology*, David W. Kraeuter, 2009.
16. *25 Years of Electronic Reviews,* David W. Kraeuter, 2011.
17. *The Pittsburgh Antique Radio Society at 25*, 2011.
18. *Sketches from a Life in Electronics*, Laten Fetters, 2012.
19. *Mr. Johnson Answers*, William (Bill) Johnson, Jr., 2012.
20. *An Oscillator Reader*, 2013.
21. *Meetings, Contests, Winners*, 2013.
22. *An Oscillator Reader II*, 2014.

Copies available at *www.lulu.com*

An Oscillator Reader II

THE PARS SIDE

FLOATERS

An Oscillator Reader II

Articles from
The Pittsburgh Oscillator
1996 to 2010

Ted Depto, P.E., Editor

With David W. Kraeuter

Pittsburgh Antique Radio Society Publication 22

www.lulu.com

Preface

Well, here it is! The much awaited *Oscillator Reader II*. This volume contains several of the many selected articles published under my stewardship as *Oscillator* editor for years 1996 thru 2010. This second volume is edited by PARS historian David W. Kraeuter. It is sure to grab your attention if you evoke nostalgia for those bygone days of radio marvels.

Enjoy,
Theodore P. Depto, P.E.
5 July 2014

This collection of articles from *The Pittsburgh Oscillator* is a continuation of *An Oscillator Reader*, published in 2013 and including reproductions of articles for 1987 to 1995. The present volume covers the years 1996 to 2010. During that time–The Depto Years–Ted managed to edit 800 pages of material covering hundreds of radio history facets. The following is a small sampling from those pages.

–DWK
May 16, 2014

Artwork credits: all artwork was reproduced from the pages of *The Pittsburgh Oscillator* unless noted otherwise. Front cover photo, Joe Patrick (see pages 67-73); cartoons on pages iii and 13, Karl Laurin; drawing on pages viii and 87, DWK; TV boosters on pages 64, 66, and 75 from *Pittsburgh Oscillator*, September 2007, pages 7, 8; photo on page 86 from *Pittsburgh Oscillator*, September 2007, page 4; back cover photo, Karl Laurin.

Contents

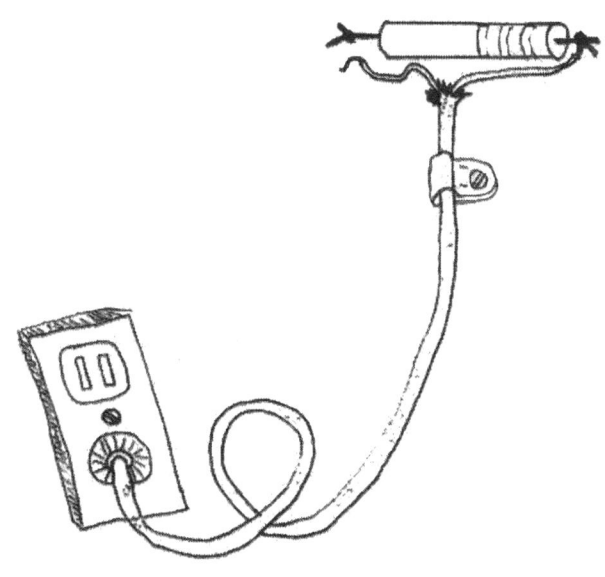

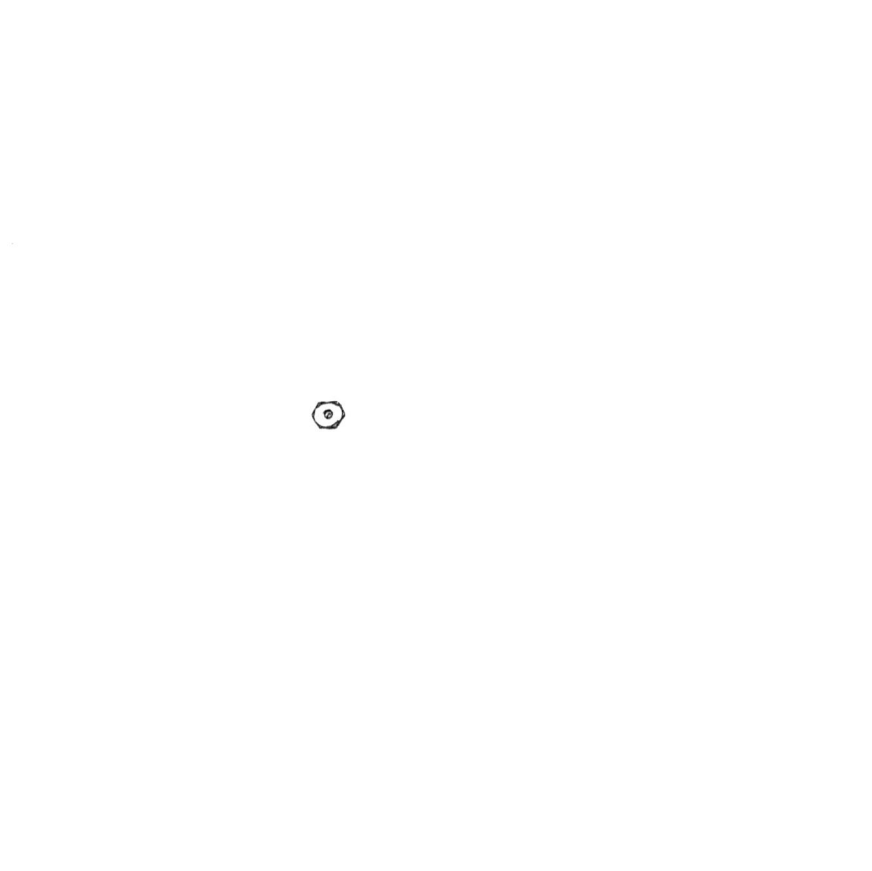

Zenith Transoceanic 1L6's
Paul R. Hyman

There are many difficulties that beset the tube type Zenith transoceanic collectors (miniature), the chief of which is the pentagrid converter, the 1L6. This type now is scarce to nonexistent. There is one exception, and that is the Tempe, Arizona supplier that cornered a large quantity of military 1L6's. The downside is that they are going for $30.00. The supply one day will be completely exhausted.

The April 1998 *QST* article titled "Zenith Goes Home" is a good discussion of the restoration of a G500. However, in the footnotes of the article, there was a web site which I explorer and found to be critical in expanding the understanding of the 1L6 and its relation to the transoceanic. The website is that of an individual named A. Padgett Petersen, PE. It has an excellent discussion of the 1L6 and some solutions for replacement. One of the more interesting solutions was the solid state equivalent.

The address for Mr. Padgett's website is:

http://padgett.performanceresearch.us/radio/1l6.htm.

I strongly recommend investigation of this website for TO aficionados.

PENTAGRID CONVERTER

1L6

Miniature type used in low-drain battery-operated receivers. Outline 11, OUTLINES SECTION. Tube requires miniature seven-contact socket and may be mounted in any position. Filament volts (dc), 1.4; amperes, 0.05. Typical operation as converter: plate and grid-No.2 volts, 90 (110 *max*); grids-No.3-and-No.5 supply volts, 110 *max*; grids-No.3-and-No.5 volts, 45 (65 *max*); grid-No.4 volts, 0; grid-No.1 resistor, 0.2 megohm; plate resistance (approx.), 0.65 megohm; plate ma., 0.5; grids-No.3-and-No.5 ma., 0.6; grid-No.2 ma., 1.2; grid-No.1 ma., 0.035; total cathode ma., 2.35 (4 *max*); conversion transconductance, 300 μmhos. This type is used principally for renewal purposes.

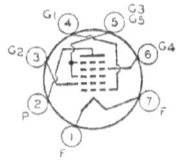

From the *RCA Receiving Tube Manual*, RCA, 1960.

On the Road With John and Sara
John W. Haught

Because of the long, cold winter, we decided it was time to warm our feet in southern climes so our destination was Charlotte, NC, to attend our first antique radio meet of 1997.

The 21st Annual Spring Meet In the Carolinas, March 21st and 22nd, hosted by the Carolinas Chapter of the Antique Wireless Association was a huge success with a large, enthusiastic, collecting crowd, with approximately 120 vendors in attendance.

The location at the Sheraton Airport Plaza, Charlotte, was easy to find and accommodations were ideal for this meet. The flea market was located at the rear, remote from the hotel activities–almost a must for a successful meet. Room rates were reduced for the meet attendees, and all meals were available at the Hotel.

Early birds could pre-register on Thursday, and the flea market was opened on Friday morning at the countdown of 8:00 AM; there was a mad rush for the vendors to remove and display their equipment, and also a mad rush for the attending buyers to run from vehicle to vehicle to be the first to purchase the "goodies". This I disagree with because after about 2 hours of rushing to see what is available, it appears like the meet is over; and we believe it hurts the attendance at the other activities of the meet.

Friday was also check-in date for the auction, limited to the first 100 items, which was held at 7:30 PM. Also on Friday was a restoration forum for the interested attendee. Saturday a trip was available for the ladies to the historic Cannon Village, the home of Fieldcrest Cannon & Cannon Mills, shopping, and a tour of a Textile Museum. This was an interesting tour for the ladies, lasting from 9:00 AM to 2:00 PM.

An equipment contest was held on Saturday, with many interesting units being judged; this gives one a great chance to observe/study some of the best in the collecting field. This to me, is the best activity at all of the meets as it gives one a chance to see and record the history of radio.

Also on Friday at 1:30 PM Ron Lawrence, President, called the

2

meeting to order, and at 2:00 PM Robert Lozier presented the Contest Wrap-up. On the agenda later in the afternoon was "Restoration of 1920's Cone Speakers" by Buford Chidester, and a presentation by John Meredith on the brief history of "E. H. Scott: The Man, His Company and his Radios." All excellent presentations.

Sara and I have attended previous "Carolinas AWA Chapter Meets" and believe this to be the best. We highly recommend attending this annual meet as the officers and attendees are extremely friendly and helpful, and it is also a good time to go South to see spring being ushered in a little ahead of our spring season up North. Hope to see you there next year!

John and Sara Haught

The Majestic that Was
David C. Ivarson, W3WBE

It was a beautiful summer day long ago …. I was about seven or eight years old and very much interested in radio. And I have an advantage over some of my peers. I knew how radios work. It wasn't very complicated—a radio works by tubes, and there were some wires underneath to connect them. You turn on the radio and tune in a station. If a tube was removed, the radio stopped working. The theory was self-evident to me.

Now in my grandmother's living room there was an ugly old Majestic radio. This relic stood up on legs and had a very plain front with a small dark brass window surrounding a dial that had an amber glow when the set was turned on. Compared to the Zenith we have at home with a nice big black dial, the Majestic was clearly inferior. Common sense dictated that a bigger dial would bring in more stations!

Emerging from the back of the Majestic was a blue wire about 18 inches long. One day I grabbed hold of its end while the set was on, and the station got louder. Thus I learned how to make radios work better. Just find the blue wire! I took this experiment a step further by touching a spoon to the blue wire. That caused such a commotion in the speaker that I was just admonished by my elders not to fool around in the back of the radio anymore. Still, I was curious. I hatched a plan to continue this experiment while everyone was out shopping the next day. What follows here is a most frightening 'home alone' story.

The logic was simple. Recalling the spoon incident, if some metal connected to that blue wire was good then more metal had to be better. Well, there was a heat register in the floor a few feet away from the Majestic. All I needed was about four feet of wire to bridge the gap between the heat register and the blue wire. A quick swing through the basement netted an old piece of lamp cord that would do the job.

With the Majestic tuned to KDKA, I twisted the end of the lamp cord to the blue wire and the station got louder, reinforcing my theory. Now it was time for the major product upgrade. As I touched the other

4

end of the lamp cord to the heat register I received a nasty shock. At the same time a loud hum screamed from the speaker. In that same split-second a dull BOOM came from somewhere deep in the bowels of the Majestic, and the dial lamp went out. The old soldier had given up the ghost! There was also a burnt smell hovering about the cabinet, and I opened more windows to ventilate the room before anyone came home.

Later that evening the Majestic was switched on for the nightly ritual of listing to the news with Fulton Lewis, Jr., but all was silent. What went wrong with my theory? For a while I entertained another theory that KDKA could broadcast a signal that would detonate old radios, but I knew that one would be a hard sell. The next day I fessed up to my awful deed. As a result, I was grounded for a few days. I now suspect that a transformer winding in the old Majestic had been grounded for a long time.

Restoration Corner
Paul R. Hyman

Once FM became available after the war, Zenith made a line of unforgettable AM/FM receivers. Early on, when the frequency of the FM Service was hanging in the balance between the Armstrong Frequency and the RCA allocation of 88-108 MHertz (then Megacycles), one of the first post-war FM receivers Zenith produced had three bands, Armstrong's (41-49 MHZ), RCA's (88-108) and the standard AM band of 550-1600 kilocycles (now KHz).

As receiver design progressed through the 50's and 60's, there were a quantity of designs which used the following tube lineup: 6BJ6 or 12BA6 for an RF amplifier, 12AT7 converter and oscillator, 12BA6 for IF amplifiers, a 12AU6 for a limiter or discriminator, and a 19T8 as diode detector as well as the high gain triode audio amplifier feeding a 35C5 power amplifier. When the Automatic Frequency Control (AFC) models came out, they used a 6KAB4 as a reactance tube and a 6BJ6 as the RF amplifier. Sets without AFC used a standard 12BA6.

One of these early gems was obtained at a flea market. This model was very early since it had no AFC.

The common faults which are almost universally found in these receivers is repaired by replacing the three-section filter capacitor, as well as the selenium rectifier.

Since seleniums are difficult to find in reliable condition you must and I did replace with a silicon rectifier. Having low forward to reverse resistance the silicon rectifier will raise the operating voltage as much as 10 volts so a series dropping resistor is inserted into the line to give the required voltage of 140 volts.

For all intents and purposes, we should have had a fully functional receiver after this work but we still had a fault that I thought was interesting enough to describe.

Troubleshooting

As the volume control was rotated through its range, the control operated normally until the last 20 per cent of rotation when the radio became silent. The 19T8 and the 35C5 tested OK and new old stock tubes did not correct the problem. I jumped to the conclusion that the fault lay in the volume control and that it had a break internally. I reasoned that if it had a discontinuity in the carbon film, and if I passed that point, then the radio would become silent. This was easily checked by measuring the resistance with an ohmmeter from the high end of the suspected discontinuity of the potentiometer and the wiper.

There was no discontinuity. The audio output is tapped from the discriminator circuit at the cathode of one of the diodes in the 19T8. At this point, there was the audio output and a minus 8 volt DC offset on the cathode. This was coupled through a 470 K resistor and shunted with a low picofarad capacitor which bypassed the RF. This went directly to the high end of the volume control. Minus 7 volts was the reading at this point. At low volume the wiper DC voltage was zero, but as rotation increased, the volume also increased, and the voltage on the wiper rose to minus 7 volts.

The wiper was connected by an .01 capacitor to the grid of the 19T8. As the control was rotated and the volume increased the voltage on the grid side of the capacitor rose to minus 7 volts also. At about minus 5 volts on the grid of the 19T8 the triode was driven into cutoff, hence the silent receiver. Replacing the shorted 0.01 tubular capacitor now had the radio playing at peak volume, and with all the sweetness that this design and speaker could muster.

Radio Days
Jason Togyer

Forest Hills combines recreation, broadcasting.

A picnic grove, a barbecue pit, two baseball diamonds and seven acres of trees weren't part of a typical Westinghouse plant. Yet that's what Forest Hills will gain as CBS divested itself of its surplus Westinghouse property. The company is about to give the borough a lodge and more than 13 acres next to Forest Woodlawn Park worth more than $500,000.

But the Westinghouse Recreation Center on Greensburg Pike isn't any old patch of grass. Radio buffs view it as the birthplace of broadcasting. The lodge, now used as a conference center, was home to KDKA's first permanent transmitter in the 1920's. It's where engineers sent the first shortwave broadcast around the world and tested early television cameras.

"There's a lot of world-important history that occurred on that site," said Rick Harris of Forest Hills, the driving force behind a nonprofit group called the National Museum of Broadcasting. He serves as its treasurer.

He and others want to use the site to display artifacts from the days when hobbyists plucked feeble signals from the air with home-made crystal sets.

Forest Hills sees the center as a chance to acquire badly needed recreation space–and maybe to generate some revenue.

"It's probably one of the biggest gifts the borough has ever received," said Borough Council President Ray Heller, Jr.

The center was one of several built near Westinghouse facilities around the country, said G. Reynolds "Renny" Clark of CBS, Inc., in Pittsburgh.

The grove at the Waltz Mill plant in Westmoreland County remains in use. "A grove in Sharon, Mercer County was sold to a developer," Clark said–a sale that left the town bitter.

Clark said CBS didn't want that to happen in Forest Hills where Westinghouse built the world's first atom smasher, and where many residents once worked at the Churchill research laboratory and the old East Pittsburgh plant.

The borough will lose $3,000 in property taxes over the site, and $24,750 has been budgeted to maintain the donated property next year. Borough manager Richard Branzel said Forest Hills hopes to make up the difference in rentals.

"I think it has tremendous potential," Branzel said. "Where can you go to rent a facility that can accommodate 500 or 600 people for a picnic? Plus, it's set up inside for corporate meetings and seminars."

Museum planners hoped Westinghouse would give them the property. Harris said they would receive up to one million dollars in state funds if they had a matching donation, and the center could have qualified.

Clark said CBS feared that the center's upkeep would overwhelm the museum group, but the borough has the resources to handle the extra responsibility.

The museum's long-term vision includes relocating to Forest Hills the house and garage in Wilkinsburg were Westinghouse engineer Frank Conrad launched KDKA's former forerunner and experimental station 8XK. The structures are for sale and could be torn down. Television pioneer Bill Brandt, now retired from a career that included work at KDKA and WJAS, can't understand why the public hasn't embraced a radio museum. "It seems to me something should have been done a long time ago," said Brandt of Monroeville, who sits on the museum's board. "People don't realize that everything that we know today, all of the communications we have, sprang from that little idea that happened in Pittsburgh."

"We haven't had the publicity we need ," added board member Bill Beal of Edgewood , who created the first local TV newscast, "Pitt Parade" in the '40's.

Heller said the borough wants to accommodate the radio museum, but fears the group's plans are to grand for the neighborhood.

"I don't think the property can accommodate any large-scale development," he said. "I'd have to look long and hard before I said 'let's put a big tourist attraction there.'"

Still Heller and Mayor Ken Gormley are confident the borough and the radio group share common interests. "The museum could use part of the lodge for exhibits or storage," Gormley said, "and the borough would likely allow it to reconstruct the Conrad garage at the site."

"There's still a lot of room for working with the community," Harris said.

The Westinghouse Lodge (about 2000).

The Coming Deluge: A Brief Debate
Don Merz, N3RHT and Jim Garland, W8ZR

POINT: Will There Be A Radio Surplus Over The Next 10 Years?
YES, says Don Merz, N3RHT

A recent article in our local newspaper caught my eye because it kind of confirms what I have been thinking about collecting radios. The article was about the accumulation and distribution of wealth in the U.S. The paragraph that grabbed me was:

> *Demographers predict that with the deaths expected over the coming decade, the next 10 years will set a record, with perhaps six trillion dollars to eight trillion dollars in estates passing through the inheritance processes.*

This reflects my thinking for sure. For some time I have been telling anyone who will listen that we are about to be inundated with hollow state radio gear. Retirees moving south will not want to take it with them and heirs will not want it–period. Unless this hobby suddenly starts attracting a large number of new adherents, there is no way that values and prices can possibly take the strain of what is coming.

I collect communication radios (affectionately known as "boatanchors" or "BAs") almost exclusively and I'm afraid that we are in for another wave of boatanchor landfill. Those few of us who collect BA's will become very particular because the choices will be many and the prices low. I expect that a lot of gear will not find any buyers and end up at the curb.

Price-wise, I am guessing that even premium brands like Collins and Johnson will be affected. The reason is simple: they were built so well that, relatively speaking, more of them have survived.

And the huge stocks of homebrew parts, receiving tubes, test gear, etc. in the basements of older hams across the land have a bleak future at best. How many more VTVMs do you want in your shack? Even the folks who might use some of the parts from modern-day projects don't want our kind of parts. My oldest son came looking for

11

some resisters the other day and wasn't in the workroom two minutes before he asked, "don't you have any quarter-watt resistors?" And the truth is that I don't have any–they are useless to me. And my ½ -watt and two-watt jobs are too big to be useful to him and his computer-controlled robot.

The ancient Chinese curse is coming for sure–we are definitely headed for interesting times.

–Don Merz

COUNTERPOINT: Will There Be A Radio Surplus Over The Next 10 Years?
NO, says Jim Garland, W8ZR

What a grim picture you paint, Don! Let me offer a somewhat more optimistic scenario. There is great interest now in collectibles of all sorts, with more and more Americans each day scouring flea markets and garage sales for relics of earlier eras. I have a friend, for instance, who collects muffateers, which is something I'd never even heard of, much less thought of collecting. She spends thousands of dollars on the things. EBay, of course, has made the collectible market much more efficient.

I'd guess the majority of collectors are not acquiring items that they remember from youth. For example, many collectors of WWII Nazi memorabilia were born well after the fall of the Third Reich. They just have an historical interest in the subject.

And so it will be also with vacuum tube shortwave gear. Obviously, the number of these radios in existence will only decrease as the years go by. Those of lesser value and less-than-first-rate condition will probably dwindle away, since the cost of time and money to restore them cannot be justified by their collector value. Many will, as you predict, undoubtedly end up on the curb.

But those well-designed radios that are maintained in good cosmetic condition will likely only appreciate in the long term. I read recently of a 1952 baseball card that sold for $83,000 to a collector. We've still got a ways to go before our hobby reaches that level! And that's not to mention the collectors who pay $1M+ for rare stamps.

In one sense, it's still too early for boatanchor collecting to become fashionable. People are still too familiar with vacuum tube technology, and many vacuum tube radios are still in active service. But 20 or 30 years from now, when the majority of collectors will have had no personal experience with the vacuum tube era, such radios will be considered true antiques. The SX-88 that brings $3000 today may be worth 20 times that or more in future years.

I expect there to be short term fluctuations in boatanchor prices. Right now, Collins gear seems to be rising in value, with Hallicrafters moving up quickly (at least for the rarer radios). They may very well peak in value and then fall a bit. But in the long term, with personal incomes rising and the supply of collectible radios diminishing, prices have nowhere to go but up. There are more than 300,000,000 people in the United States today, and many will be moving into the age bracket over the next decade when they'll likely become interested in collecting. Their numbers will easily absorb the supply of radios appearing at the estate sales of us old codgers.

Should I Plug in the Old Radio I Found at the Flea Market?
Lou Gaetano

Ihave been telling people that when you find old radios at flea markets and estate sales, you should never plug them in or turn them on to try them out. Look to see if items are nearly complete. If a radio is in complete enough condition to allow restoration, a piece that you are interested in and at the right price, buy it and get someone to inspect it before applying power.

First a good safety inspection should be performed to verify that there is insulation on the primary winding, no visible shorts, etc. A quick check with the ohmmeter to verify that nothing is shorted doesn't hurt either. Power should be applied gradually (as using a variable transformer, increasing voltage over a period of a day or longer.) This will allow moisture to be driven from power transformers by gradual warming before the full voltage is applied. If possible, (but not for series string units) the rectifier tube should be removed for the initial application of power. It also helps to first separately bring up the B+ voltage to reform filter capacitors with tube filaments out and using an external power supply. If you do this many radios will work afterward with little or no repair! At least this will prevent further damage, flames, or personal injury.

A couple of days ago, I came across a perfect example of why to do the above. One of the radio club members gave me an old late 1940's "coin operated" radio to repair. At first look, other than being rather dirty, it didn't look like there would be much wrong with it. A safety inspection proved that it could have been disastrous to plug this in! The radio was equipped with a sheet metal case. There was a slot in the coin operated timer unit for the radio power cord to go through. Someone had closed the door on the cord without placing it in the slot. Needless to say, the bare conductors of the damaged wire were making contact with the metal chassis. Past this, I inspected the assembly further. As designed this radio had wooden insulators under the chassis.

The radio was mounted with shoulder washers to insulate the radio chassis from the metal case. The receiver is a series filament type without a power transformer–HOT CHASSIS design. That's right, one side of the power cord goes right to the chassis. Someone that previously worked on the set had thrown away one of the wooden insulators and replaced some missing insulating shoulder washers with regular metal hardware. Now, even with the power cord repaired, the sheet metal case would still be now-electrically connected to one side of the line. After disassembly, trying to charge the filter capacitor, I found that it was shorted. A quick electrical check found that the audio output transformer was also shorted to ground. Believe it or not, all the tubes were good. The radio will work again, but some precautions prevented further damage.

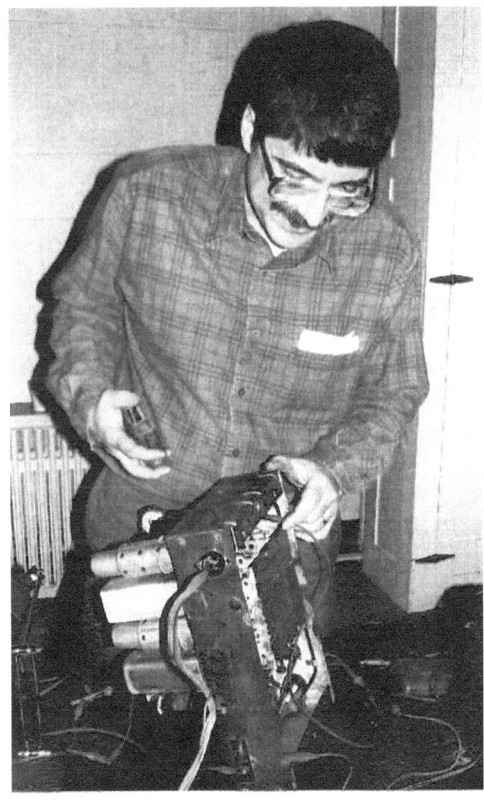

Lou Gaetano at the March 2005 PARS meet, which featured our first radio service clinic.

The Philco Model 41-608H
Ted Depto, P.E.

Reconstruction of a Grand Old Classic

I collect Philcos mostly because they are excellent performers and in many ways superior to other brands even though they were not always known for their exquisite cabinet, or followed the trends of the day. This leads to a radio that I was able to acquire recently, known as the Model 608H, manufactured in 1941. A large console with a pull down front door which also houses a 12-inch electrodynamic speaker plus the phonograph, it is a nine-tube loctal based superhet with two type 41's in push-pull output.

Beam Of Light

Although not considered by many collectors, this particular model has two things going for it. It incorporates the famous "Beam Of Light" reproducer in the phonograph section, one that caused Philco engineers to have second thoughts about the system and whose use was limited to a scant two-year production run. (A tiny bulb located under the tone arm of the changer uses oscillations generated in the radio and focused on a small mirror whose light variations from the moving stylus energize a photocell, and as the stylus follows the record grooves the modulated beam translates into a small voltage which in turn is amplified conventionally.)

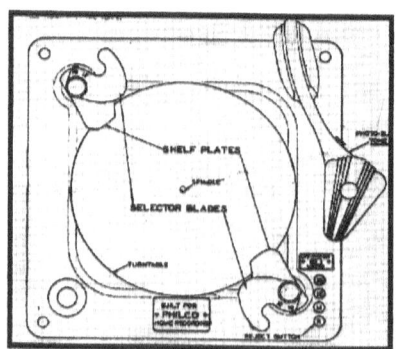

Drawing of phonograph
shows tone arm with end
enlarged to accommodate the
Beam of Light reproducer.

Secondly, this particular model also incorporates the provision for an early home recording system and with the accompanying microphone which I found to be with the radio, you can make your own recordings on wax or plasticized disks. When I first saw the set sitting in the corner of a damp basement, I really had second thoughts about my ability to restore it. The bottom moldings were rotted and the set was leaning to one side because the rear moldings had partially disappeared. But being a Philco collector and never having had the opportunity to find another in the last four or five years, I decided to take a chance.

Woodwork

After carefully removing the chassis, speaker and turntable from the cabinet (the phonograph plate completely fell apart from its glued position), I turned the cabinet upside down to begin work on the moldings. Several of the struts used to hold the cabinet in position were rotted, as were the mounting board for the Philco loop antenna. Veneer was also missing from the phono mounting board, but surprisingly, the chassis and the exterior veneer on the cabinet were in good overall condition, albeit loose from glue failure. Using a bandsaw and a sander, I fashioned two new bottom moldings (the straight sections) made from white Ash, fastened with glue and two-inch wood screws.

Spraying lacquer sanding sealer and toner in the original Van Dyke brown, this finished one part of the restoration. The grille cloth was intact but soiled and when I tried to wash it, it shrank to the extent that it didn't cover the front area as before. This prompted a call to John Okolowicz who handles grille cloth for a number of older radio styles. I was able to find a close replacement.

Turning my attention to the radio chassis, I replaced the filters and line cord, and sprayed the contacts on the band switch and the volume and tone controls, and peaked up the I.F.'s for maximum. As the restoration progressed I became more enthusiastic in the completion since things were starting to fall in place and because I needed the room that this radio was taking up. I hadn't yet begun to tackle the Beam of Light phono system.   [To be continued.]

Converting a Battery Tombstone to AC
Ted Depto, P.E.

I've got this huge Arvin tombstone from 1936, a model 617B, and it has been sitting on the shelf looking back at me for all of the ten years that I have owned it. All the while very quietly collecting dust. I really never entertained the thought of firing it up since as a static display, it fits the bill for large '30's farm sets. Its condition is quite original, missing only one or two knobs, which does not seriously detract from its interest.

It is always nice to have a static display but in this case, and after much deliberation, I finally decided I would rather have a functional radio (note–purists, patience please) and in this regard the vibrator powered radio just is not practical.

There are several things you can do if you want to make it operational, namely connect it to a heavy duty six-volt power supply, add tons of filtering to reduce the hash or carry a six-volt storage battery around with you, or in my case, which I finally decided upon, convert the radio to regular 110-volt AC house lighting circuits.

Since I chose the latter, the criteria I used in making this conversion was threefold. 1. Not to modify the chassis hardware other than adding tube sockets and power transformer. 2. Keep all wiring except for the AC needs as original. 3. Use the original speaker and output tube working electrical parameters. I found this system to be most rewarding since Arvin made a model 617 which was AC line operated, so I felt some license to operate on the radio.

The Conversion

In order to effect the conversion, the large transformer and vibrator were removed from inside the vibrator compartment on the chassis along with the aftermarket filters which were installed under the chassis by a previous radioman.

Filter capacitors are always a problem, so no matter what type of radio you are restoring, new ones should be installed on the chassis at the outset. If you notice the wiring diagram for the Arvin, the 6.3-volt winding for the filaments have the proper voltage, but AC on the 19 output tube filament and its counterpart 15 in the RF stage would present a hum problem if not addressed. And it would not be very easy to remedy. (Once I worked on a Zenith tombstone farm set and to make it perform as intended, it required upwards of 4,000 mfd. filtering to reduce hum to a negligible level).

So there you have it. Easy conversion but maybe not. I located a small power transformer which I calculated should handle the current drain very nicely, and mounted this Thordarson on some of the same mounting holes as the original. This transformer had a 175-0-175 high voltage winding, center tapped, along with the usual 5.0 rectifier heater winding and the 6.3 volt winding to power up the rest of the set. Next the vibrator retaining ring was removed and all wiring unsoldered from the five-pin synchronous vibrator socket. As a precaution, I tested the wafer socket for high resistance shorts because carbonization can be a problem in these old sets. So far so good.

I added an octal socket, used a 5Y3 for rectification adding approximately 1000 mfd. filtering to the heater winding after installing a bridge rectifier and a 22-ohm dropping resistor to bring the filament voltage to specification. I used a 25-watt potentiometer in addition to the new filters in the B+ line to adjust the voltage to the plates of the tubes, and after setting the voltage at 175 volts, removed the pot and measured the resistance with an ohmmeter. I then installed a five-watt fixed resistor to clamp the voltage at this level. After trimming up the IFs and installing four new dial lamps, the radio plays as it should with a slight but tolerable hum level.

Arvin used metal cased capacitors mounted on the chassis, and to keep the original look, I disconnected the wiring and added new orange drop capacitors underneath. The original radio must have required a heavy current drain battery from the amount of dial lamps (seven) in the radio and its corresponding six tube lineup. Some of the original resistors had changed value so any that measured more than 20% off value I replaced (about four) and finally located a drive belt to operate the tuning of the radio.

6.3 Volt Bridge Circuit

The DC supply for the filaments and heaters in the radio is a straightforward design taken from schematics of other vintage radios and not unlike a small universal power supply. As heavy current is drawn thru a DC supply of this type, ripple level will increase to the point where hum is a problem. And to address this condition, massive filtering with an additional choke is needed.

The addition of a choke in a T network of filters instead of using a high-wattage resistor will not only substantially reduce the amount of ripple at higher current levels, but has the advantage of less voltage drop. Since the series-parallel heaters in the Arvin did not have to be rewired it made for an easier conversion. A wirewound resistor had to be inserted into the A+ line to keep the voltage at approximately 6.2 volts even though the radio operated as well on 6.6 volts in preliminary tests.

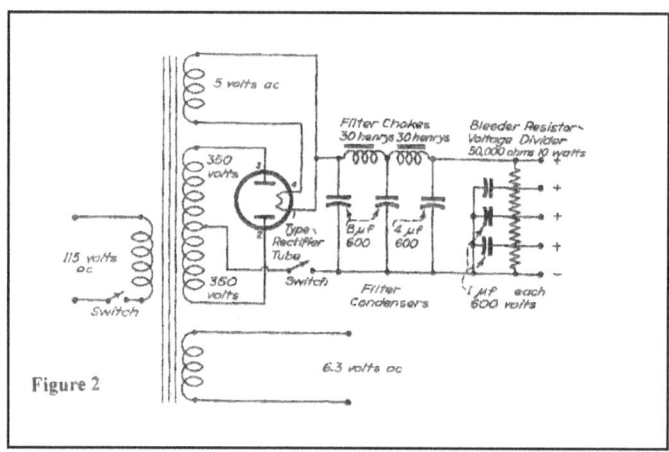

Figure 2

The actual design of the power supply is similar to the one shown in Figure 2. Although this layout shows a separate filament transformer it was incorporated as part of the main power transformer in my conversion. Necessary in this project.

The final A supply is shown in figure 3, and does incorporate heavy filtering with the addition of a filter choke. Now the Arvin is a functional radio and aside from the wiring changes still retains the old charm, look and feel, minus the vibrator hum.

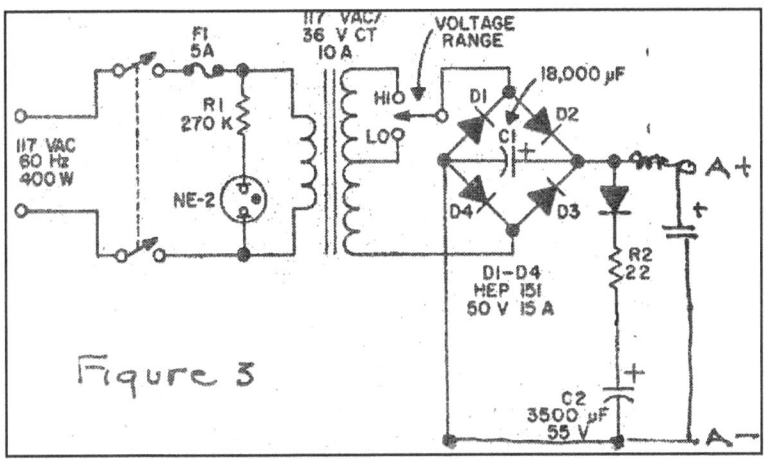

Figure 3

I thought you might be interested in this conversion since we all have old battery radios that just beg to return to life after a hiatus on the shelf. Good luck on your next restoration job.

More information and pictures at:
www.radiomuseum.org/r/arvin_617b.html

RHYTHM MAID 617
6 TUBES $59.95
1936

The Midwest Story
Ted Depto, P.E. and Mike Simpson

Midwest Gets 'Em Coast to Coast

Part I
Forward

In the early '20's retailers Montgomery Ward and Sears-Roebuck were giants among pioneers in mail order sales of radios. Powell Crosley, based in Cincinnati, Ohio, was selling Crosley Harkos around this time thru his dealers, and he was well on his way to becoming a force in this new industry. Atwater-Kent was firmly entrenched in Philadelphia producing quality products and RCA was the outlet for Westinghouse, GE and others.

This was a time when many major manufacturers recognized the potential of radio and jumped into the market. Companies such as Airway, (the Airway Vacuum Cleaner Company), for example, diversified by building radios, a small step for a major industrialist to enter the radio market since the facilities for producing their patented vacuum cleaners were existing and retooling for radio production was a small step indeed.

Enter Midwest

As a corollary, another manufacturer was busily engaged in selling radios.

Midwest entered the picture in 1920 by one A. G. Hoffman, first producing radios as a hobby in his home at 404 Dury Avenue in Cincinnati, and later using direct marketing thru mail order. Midwest quickly became a household name in the business. Calling itself "Miraco", short for "The Midwest Radio Company", later reincorporated as "The Midwest Radio and Television Corporation", the company produced lower cost radios emulating higher priced sets, and in many respects duplicated them in performance. Midwest was not to be taken lightly through the '20's. It had the backing of a large Ohio bank which guaranteed funds for research, manufacturing and advertising. It is known from Midwest literature that Midwest in later days manufactured its own coils, transformers and speakers. A lot of other manufacturers simply assembled sets from parts from other suppliers.

How It Began

The earliest information on the company that I could find indicates that as Midwest began to expand it outlived the space provided, and records show that in 1923 it moved to 812 Main Street near the business hub. Eventually Midwest occupied a whole block on 8th Street to build its products.

I really never thought too much about the Midwest brand until one day last January when I happened to meet Mike Simpson (my co-author here), a noted Midwest collector and a member of SCARS, the Southern California Antique Radio Society.

Most radio collectors jump at the chance to own a Scott All-Wave, a Silver Marshall or even a large Philco such as the magnificent 38-690. Mike told me that since virtually everybody had the large multi-tube models he decided to collect a brand that, although widely advertised in radio magazines, was mostly looked down upon by collectors for the many tube chassis of which some tubes appeared to do nothing to improve performance. (It might surprise you but some Scotts and others will also work with tubes removed but at a lowered performance). Mike also reiterated that he decided on Midwest partially because of the foregoing and the fact that he was able to purchase one as a teenage, and we all know about nostalgia.

Mike settled on becoming a Midwest collector and now has over 50 models. One of their best selling radios in 1924 was a three tube model R-3, as shown in the advertisement reproduced here.

Probably the best selling model in the '20's turned out to be the Miraco Ultra 5 in a table model shown here and although it could be ordered as a console with a matching speaker stand, most radios sold were the walnut cased table models.

During the boom and bust of the '20's, Midwest fared rather well as radio manufacturers go, and started its ascent into higher tube count models in the early '30's. It is known that the cabinets were sometimes simply nailed and glued together and lacked the craftsmanship and joinery that other manufacturers relied upon.

Undaunted, Midwest continued to market additional models with battery powered sets, smaller AC table models and finally the large consoles with auditorium type speakers. Ads from 1935 show that Midwest had the D-18, a fifteen tube console using a 15 inch pedestal speaker. It is housed in a walnut cabinet of traditional design with the characteristic Midwest chassis.

Midwest advertised very heavily in the rural market and began offering AC packs in 1927, but continued to market battery sets through 1932, 1933 and 1934. Some of the models are very collectable today, and command high prices in the radio market.

But to return to the advertised models for 1927, many publications saw an advertisement for the "light

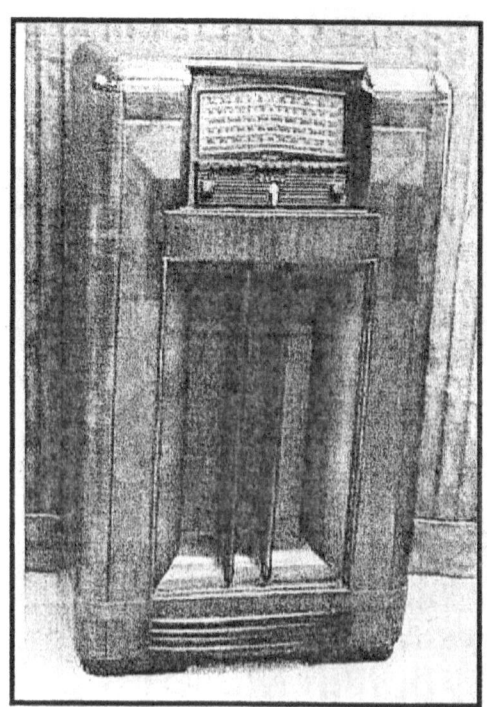

Midwest Model D-18

socket units", which essentially upgraded even their old battery models . Some Midwest models carried over from one model year to another (probably to get rid of the existing unsold models in stock), but Midwest really came into its own during the 1935 season with the Dreadnaught chassis and V-front styling. There is no mistaking a Midwest console radio.

Midwest 18-36 Dreadnaught Chassis V front styling DD-18

As the Midwest designs improved, sales also did, and in the '20's Midwest was an established mail order business. Not all of their models of the '20's are included in the photos.

Part II

L ast time we discussed the origins of the company (the company in its literature cites 1920 as the date of commercial radio manufacturing for the public.)

Beginning about 1927, when the three dial tuning arrangement of most radios was beginning to fade, Midwest introduced the "Unitune" single control, a marketing ploy with the three tuning capacitors ganged together. This proved to be a very popular change and although the electrical specifications remained essentially the same as the 1926 models, the radio-buying public wanted the latest in radio design, hence the single dial radio.

This was also a time when the AC light socket sets were prototyped and with the development of the RCA type 26 and 27 tubes specially formulated with AC heaters, the stage was now set for light socket AC radios. Not just Midwest but virtually all radio manufacturers hastily geared up for this welcome change. No more battery acid on the living room rug. Prior to this development many manufacturers including Midwest tinkered with the idea of using AC packs to convert their battery radios to eliminate the B battery for the plate supply.

The A supply was more of a stickler since high current rectifiers were not yet reliable and most supplies used tungar rectifiers with floating DC over a storage battery. Incidentally, the K-Power AC power supply built in East Pittsburgh about 1927-28 that was demonstrated at the April 2001[PARS] meeting was an example of this technology.

The company produced several small AC sets in 1933 in addition to a nine-tube (9-33) and an 11-tube chassis which could be fitted into various cabinets or supplied with Midwest cabinets. These models were designated as A series. 1934 had a large 15-tube (15-34) chassis four-band with automatic volume control and a type of bfo heretofore not found in radios other than some higher-priced sets such as Sargent-Rayment and National, to name a few.

The 1935 high tube count models put out by Midwest have several interesting features. They incorporate a Micro-tenuator– Midwest's exclusive expanding and contracting i.f. transformer which essentially compressed the response curve and was patented as no. 721240. Another, amplified Q-AVC introduced earlier, was touted in the Midwest brochures as a way of softening intercarrier white noise, a form of squelch.

Heavy duty power packs, consisting of the electrolytic filters, chokes and power transformer were more rugged than before using better insulation. The 16-tube all-wave covered the frequency range of 33 megacycles (MHZ) to 125 kilocycles (kHz).

Models A, B, D, K and PR were the mainstay of the 1935 Midwest line of receivers, with the PR-16 called "The World's Finest Musical Combination . . . the Midwest Super Deluxe Radio Phonograph!" This radio phonograph was massive in size, rivaling the Zenith Stratosphere. Its dimensions are 49 inches high, 29 inches wide and 19 ½ inches deep.

Those Midwest models paved the way for even more exciting news from the company as the 1936 models would see even more features (and hype) at an affordable price to attract the savings-conscious buyer.

In the console line, 1936 saw the introduction of the "dreadnaught chassis" which would be marketed for several years with only minor changes in tube lineup, control features and cabinet styles. Inspection of this particular chassis reveals it does in fact utilize all the tube types for a particular purpose, and refutes the critics as saying some tubes do nothing but light up.

It is true, however, that many of the same tube types (6SN7's for example), are used in push-push and push-pull arrangements or in parallel circuits fed by dual triodes when essentially a pentode would give the same result. We know that high fidelity sound quality was at this time in its infancy and perhaps the triodes operating as Class A drivers would give a cleaner sound to the large 15-inch electrodynamic speaker employed in Midwest's top of the line models. Note that almost all manufacturers have by this time abandoned the small window dial in favor of a more elaborate system of dials behind glass with airplane dials, or telephone type. Some (Zenith) used flywheel tuning to quickly tune across the bands. In keeping with the low offered price, Midwests did not come with a dial

THE NATIONS PARADE BEFORE
YOU... *When You Turn the Dial of A New Midwest!*

From Midwest ads.

glass but instead an interesting finger wheel assisted in tuning the multi-band chassis with its many gradations.

The H-5 tombstone proved to be a good seller as well as a stellar performer, still using only five tubes in its circuitry. Midwest stated, "Do not confuse this finer Midwest radio with the 'pee-wee' cigar box models or ordinary small midgets that sell at bargain prices. They all bring in voices, speeches, music . . . to be sure . . . but, like in a telephone, the full tones are ordinarily missing. This Midwest radio is a finished musical instrument."

With only $5.00 down this radio could be delivered to your door with low monthly payments, and while Midwest derided the low-priced models of other manufacturers, it too was involved in this low priced marketing war of words reaching out to customers.

Part III

With the multi-tube models advertised heavily and the fact that most manufacturers were competing for a share of the lucrative high-end market (profits on some models exceeded 35%), Midwest indeed made a significant dent in the sales picture overall, outselling Parmak and Colonial during the late '30's.

1937

The 1937 models of Midwest included table models L-11, D-7, K-11, a 5-band set (the number after the model letter designation indicates the number of tube in the receiver), consoles P-14 with 14 tubes, E-7, a tube three band console, S-16, 5 bands, T-16, five bands and model W-18, X-18 and the six band radio phono Z-18. This particular radio phonograph used a changer which literally flung the spent record into a plush lined compartment on the left side of the cabinet.

1938

1938 models BB-7, CC-8, GG-9, LL-12 and LL-10 were the table and tombstone models offered (I have yet to find a Midwest cathedral style radio although by 1938 the era of the cathedral was about over). In this year console models TT-18 with motor drive, the Regal

Victoria, model SS-18 with motor drive and the double door model UU-18 and YY-20 with its motor driven chassis were being advertised in the company brochures.

Remember, you could buy a Midwest chassis and speaker which could be placed in your own cabinet, but the offerings of the company by now vastly improved the number of available factory cabinets for the discerning radio fan and many buyers ordered the units complete.

1939-1941

Years 1939 and 1940 saw a decline in the production and arguably quality of Midwest models. CP-17, 12 CP-15 (with record player) and model D-15 rounded out the line but now several other manufacturers (Arvin and Wells-Garner) were teamed up with Midwest and their radios were being sold with Midwest nameplates.

One of the more notable models was the 1940 "Console Grand" with catalog number D-18. It featured motorized tuning, organ key tone control, a double action tuning eye, dual speakers and the patented Organ-Fonic tone filter. The cabinet stood a majestic 43.5 inches tall, 25 inches wide and came in toned walnut. It was also available as model D-12 with a twelve tube chassis.

The model R-8, one of the better selling models, has six bands, eight tubes, a hefty power transformer and utilizes a tuning eye. Dimensions of the cabinet are 8 ½ inches high, 13 inches wide and seven inches deep. It is housed in a two tone walnut finished cabinet.

Post War Production

Production after 1946 never really saw the glory days reached in the mid to late '30's but Midwest continued to offer a good selection of console and table models, almost all now with the new horizontal styling and featuring exotic cabinets.

Television Models

Midwest came into the television market rather late as 1950 introduced its console model KX-19 for the 1951 model year. Along with a variety of FM radio phonograph consoles (utilizing the

Midwest model KX-19 for 1951

symphony grand chassis) this was the extent of the company's products. As consumer interest in television receivers grew, it appears that Midwest did not completely rise to the challenge of large makers such as GE and RCA. Along with a host of small manufacturers during this time, Jackson, Tele-King, Muntz, Radio-Craftsman and Transvision, to name a few, the stage was set for Midwest to quietly ease from the marketplace in 1955.

Conclusion

In defense of Midwest, it might be well to remember the sometime rocky road the radio manufacturers had to ride in order to survive in this enormously competitive environment.

Not only in dealing with the recessive economy prior to World War II (radio's Golden Age?) but also in the recession that occurred during the time when millions of servicemen mustered out of the various branches of service after the war and before the post war building boom began.

The resilience of the company to turn attention from wartime production to peacetime consumer products stands as a tribute to the American ideal. From products obsolete practically by the time the designs reached the production lines, and the ever changing consumer demands, it follows that Midwest finally lost its place in the market.

Its legacy might well be described as follows: "provided a low cost avenue for thousands to enjoy the best programming on the airwaves."

References

Midwest Radio and Television Corporation brochures for 1935, 1947, 1951 and 1954.

Radio Retailing, June 1940.

Mike's Midwest Museum website:*midwestradiomuseum.com*

Radio News, February 1951.

Photos from the authors' collections.

A Catalin Saturday in Clarksville
Michael Male

It was another Saturday, and of course, as every working radio collector knows, that's shopping day, but something told me this was going to be a special day. It was that feeling you get when you know that something you've never seen before is going to show up. Like that Catalin resting in a barn for 50 years waiting for you to bring it home.

My day started with a visit to the Antique Mall in Ionia. While I was sifting through some items I had seen 100 times before, I was approached by a short man dressed in coveralls.

"You the radio man?" he asked. "I'm a radio collector," I replied. "Well, I was out by Clarksville. There's a man who's got one of those plastic radios. Ya know, in all the colors. Wants to sell it too," he said with a smile.

Trying to keep my composure and refrain from bursting out in a wild dance, I asked meekly, "Where?" Sensing my excitement, he smiled again and said,"it'll cost you $10 for the address. I would've bought it myself, but I'm not interested in radios, and $20 seemed like a lot for some old butterscotch colored radio."

I could feel my heart pick up 20 beats a minute.

"When ... when did you see it?" was all I could get out.

"Just come from there," he replied. "Maybe an hour ago."

My hand was shaking as I handed him the $10 bill. He gave me a crumpled up piece of paper and said, "The guy was just leaving. Said he won't be back until 6:00 tonight. Go there then and he'll have it for you. Good luck!" Pushing the $10 into his pocket, he waved and walked away.

My head was reeling. "Butterscotch" rolled over and over in my mind. "Fada, Garod, Addison."

I quickly checked the address before the old man could get out of sight. "2689 Hemlock Rd, Clarksville." What luck! I knew where that was. This was going to be the most profitable day ever.

I spent the rest of the day not really seeing what I was looking at and glancing at my watch frequently. I began to devise a plan. I would go at five o'clock and be there when he came home. By three o'clock, I could wait no longer. My palms had begun to sweat. I would just have to take a chance on his coming home early or wait until he did.

The drive took me a mile out of Clarksville, and to a red and white mailbox reading, "U. B. Took, 2689 Hemlock Road."

I pulled into a long dirt driveway, which wound back a half a mile to a brown shingled house. The porch was falling apart. My eyes scanned the yard. No cars. Guess I was too early. I settled back for the wait. About 15 minutes later, a blue Ford sedan came down the dusty drive. My excitement grew, as I thought my wait was over sooner than I had hoped. The sedan pulled in next to me. I waited for the driver to get out, but he just sat there looking at me. We stared at each other with blank expressions on our faces.

But soon another car came down the drive and pulled in next to us. And then another. Before long, there were some thirty cars lined up in the drive. Each driver sat staring at the others in disbelief. Looking at the house, I spied a note placed on the door. I quickly went to it and read:

Catalins, Catalins, everywhere,
So Greedy fools, please beware
Next time bargains that are too good to be
Please remember me.
*   Signed,*
*   U. B. Took*

One by one, everybody else walked up to the door to read the note. As we begin our long parade out of the driveway, everyone waved at each new car that approached the little brown house.

Yes, the old man had taken us, or maybe we had taken ourselves. Either way, it had been an interesting Saturday.

The Philco Mystery Control:
Forerunner of the TV Wireless Remote
Richard Brewster

Who would buy a TV without a wireless remote control? But long before the "remote" was commonly available on television sets, it showed up on a radio. In 1939, Philco began marketing a console radio with a wireless remote called the "mystery control."

Then why wasn't wireless remote standard on TV sets till much later in the TV revolution? Actually Zenith had a remote control system called "space command" in the 1950's. It utilized a light beam and photocell (later an ultrasonic signal) with the tones generated by a handheld mechanical clicker. It worked well as I recall. But I believe that it wasn't popular for several reasons.

Families typically watched the same channel for the whole evening–imagine that! And presumably commercials weren't nearly as annoying then as they are now, so even the mute was not considered necessary. And there were only a few channels. Some folks got just one or two. Growing up in the New York City area, we were privileged to receive seven channels even as early as 1949 when we got our first set. But I believe that the real reason for the resurgence of interest in the TV Wireless remote was cable TV. With so many channels available, there's a strong tendency to want to "channel surf." In addition, infra-red solid state controls have become very inexpensive.

Back in 1969, when I worked for Westinghouse, a small group at my location was developing an infra-red remote control system using LEDs. That, of course, is the preferred system for most of the wireless remotes now in use. Lately RF is sometimes used, echoing the system chosen by Philco in 1939.

Strangely, a Philco remote control transmitter was one of the very first items in my collection, having been saved from my childhood junk picking! Many years later I acquired and restored the Philco 116RX which it controlled. The Philco 116RX is a really neat set. The one I restored in the early seventies worked so well that I could go out into the yard, at least 100 feet away from the set, and still control the volume, change preset stations and even turn the radio off. But once off, it could not be turned back on except at the set. The nine inch by seven inch by six inch remote unit uses a type 30 tube with three-volt and 45-volt batteries and a simple oscillator circuit. The oscillator is pulsed by a telephone-type dial. These RF pulses are picked up by a loop antenna in the cabinet base and processed by an auxiliary 5 tube short-wave receiver on the main radio chassis. A stepping relay (enclosed in a large sound-insulated container) which selects the stations, is controlled by a 2A4G thyratron tube, surely unique for a home radio!

Those guys at Philco were pretty clever. Later, they came out with two of our favorite collectibles: the first battery-operated TV in 1959 and the famous Predicta.

Unfortunately, I disposed of the 116RX due to lack of space. I did save the original remote unit and a few years ago acquired another Philco 116RX that I plan to restore some day! It shall remain an unusual part of my eclectic collection.

PHILCO 116RX
WITH MYSTERY CONTROL

35

The Kemper Radio Corporation
Floyd A. Paul

Kemper Radio Beginnings

Radio Doings magazine of February 12, 1928, had an article on the Kemper Radio Lab. The following paragraphs paraphrase that article.

In the summer of 1925 Guy A. Kemper started a business manufacturing a portable radio in his garage. This radio had a chassis, a loop antenna, and a carrying case. The initial effort was truly a garage shop operation. Guy made one or two radio sets at a time, offered them to some acquaintances, and made several sales.

His early problems stemmed from selling the sets too cheaply and he wasn't making any money. He might not have survived the year of 1925 with his initial efforts, however a Harold E. Cluff came into his life. Harold had been an ore miner in Nevada, had his fill of trying to find gold, threw his pick away and headed for Los Angeles. He met up with Mr. Kemper, saw the possibilities of selling the radio that Guy was making and offered his services. Guy accepted the challenge to work with Mr. Cluff and the two of them started working together.

Guy made a few sets a day and Harold would put one set under each arm and walk Los Angeles city streets in the mornings. By noon he had always sold the two sets. He would return for more. Harold apparently had a good feel for getting the business going. He found he was selling sets faster than Guy could make

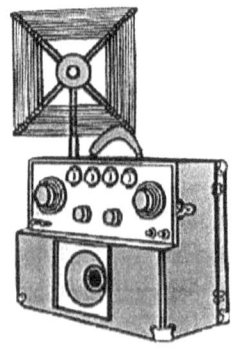

Drawing of the Kemper Super-Harmonic, the first model he made.

them. Guy then hired a new technician to assist him in making portable sets. Now there were more sets available than could be sold. They added another man to help sell them. Soon Harold became sales manager, as he hired more salesmen.

By the spring of 1927 the company became Kemper Radio Corp. Their motto had become "Direct factory to consumer no middle man."

In 1927 Mr. Percival Townsend became production manager and created the Radiomobile set in that year. The sales personnel moved to the Ambassador Hotel where all of their sets were displayed. (In a *Radio Doings* ad of April, 1927, Kemper Radio was being offered to hotel guests for a nominal rental charge.) The *Radio Doings* article concluded with the statement that the company was doing very well financially in 1928.

Kemper Sets

The first model of the Kemper portable radio was called Super-Harmonic and was made in 1924. It had four tubes and had holes in the panel to allow the tubes to be seen from the front. The second set was also called Super-Harmonic but had five tubes all mounted inside with no panel holes. By summer of 1926 the next portable which was labeled the K-5-1 was being manufactured. In the next two years the K-5-2 and the K-5-3 followed. The K-5-1 and K-5-2 had a meter in the center of the panel while the K-5-3 had a meter on the left side. The K-5-3 came in at least two styles, two thumb-wheel tuning knobs and two separate tunable condenser knobs.

Kemper Model K-5-1

The Radiomobile floor set (K-5-5) was a unique design with three-inch castor wheels, a loop antenna, and AC power supply,

Sovereign tubes, an upper and lower chassis, and an Aerofonic air column speaker (which was made in Los Angeles).

The Ampliphonic, the Duophonic and the Eletrophonic models were attempts to make a console radio out of the portable set. The company assembled tables, portable sets and AC power supplies into a console arrangement.

The SG7 model, with screen-grid tubes, came out in 1929-1930. The chassis in that model was used in the SG71 model, as well as in the grandfather clock radios. The grandfather clock radio was made in two models: with and without a phonograph. Karl Manthei of Fairfield, California, collector of grandfather clocks, told the author that Kemper grandfather clock radios were made between June and December of 1931. (See the radio set listing for an organized compilation of models and dates.)

Guy Kemper was born in 1894 in McComb, Illinois. He died in 1968 in San Louis Obispo, California. He served in WW I as a part of the 347th Field Artillery. He started the Kemper Radio Laboratories in Los Angeles in the 1924-1925 period.

Guy Kemper joined the Institute of Radio Engineers in 1927 and remained a member until 1930. By the time the company went out of business in late 1931 or 1932 Kemper built a successful business with outlets in Santa Monica, Hollywood, Pasadena, Long Beach and Los Angeles, as well as outlets in San Francisco and Seattle. His sales staff numbered some 150 representatives at one point in the 1928-1929 period.

Plant Locations

Guy A. Kemper

Mr. Kemper's first assembly site was in Los Angeles in his home at 2718 Moss Avenue. In September of 1925 the shop moved to South Hope Street. In December of 1925 the plant moved to 219-221 Venice Blvd. near Hill Street. In April of 1926 the company moved into the Kotzin Building at 1236 South Santee. In 1927 Guy Kemper moved his assembly line into another manufacturing company's facility. (This may have been Jackson-Bell or The Wireless Shop).

Financing

During the first few months of his business the company was called Kemper Radio and survived on Guy Kemper's own capital money. During the manufacturing of the first "Super-Harmonic" portable radio, investors were brought into the business. Then in 1927 the company changed its name to Kemper Radio Corporation and went public.

Tom Kemper's Recollections

Guy Kemper's son, Tom Kemper, of Northridge, California, gave the author copies of Kemper Company bulletins, photographs, and much information about his father. Tom said the remembers his father saying, "Our company made some 90,000 sets". Tom also found a National Radio Institute lesson well annotated by his father. Notes indicated he knew the lesson material well. Tom also remembers his father saying he helped build a radio transmitter in the Los Angeles basin. With this background in radio understanding Guy Kemper started his company that spanned some seven years (1924-1931).

Chronological Kemper Radio Model Listing

(Derived from Lefax handbooks, McMahon's book *Radio Collector's Guide*, and Kemper Company sales bulletins).

Year	Model/Name	Price	Description
1924-25	Super-Harmonic	$ 98.50	4 tubes, 4 holes in panel
1925-26	Super-Harmonic		5 tubes, no panel holes
1925	K-5-1	135.00	5 tubes, meter in center
1926-27	K-5-2	135.00	5 tubes meter in center
1926-27	Ampliphonic		K-5-2, small table, with AC supply, console style
1927	K-5-3	135.00	Meter on left side of panel, with two vernier knobs or two drum-dial controls
1927	K-5-4		Speaker on side, AC supply (Stew Oliver reference)
1927	Radiomobile K-5-5	225.00	4 Sovereigns, 71A, BH tube AC supply, floor set (Stew Oliver reference
1928-29	Electrophonic		

1928-29	K-5-6	99.50	5 tubes, (See McMahon's book)
1928-29	K-5-7	74.50	"Kompak," 2-26s, 27, 112A & 80 (Lefax handbook reference)
1928-29	Duophonic		K-5-3 in console arrangement
1928-29	SG7	129.50	"Moderne," 7 tubes, 2-27s, 2-24s 2-71s and 80
1930	80	69.50	"Kompak," 2-24s, 2-27s, 45 and 80, tombstone set
1930	SG71	141.50	"The Chief," console, uses SG7 chassis
1931	Grandfather Clock	375.00	SG7 chassis, with or without phonograph

Kemper Stock Offering Certificate circa 1930

41

When Ed Speaks–My First Collectable Radio
Ed McGuigan

I'm a believer in coincidence, fate, even ghosts (having been visited by the same). I got to thinking about my first radio purchase. It was about 15 or 16 years ago. There was an ad in the paper for a Silvertone table radio. I called and the lady said it was old, big and wooden. Well, that was enough for me. I got directions to a house a couple of miles from me. When I got there, I was greeted by an older woman, her family and a worn looking Silvertone tombstone. I asked the price and was told something about three times what it was worth. But hey, it was my first purchase. I knew nothing about old radios. As far as I was concerned, there might only be 100 of these left in the world. I paid the money, thinking this was worth a week's groceries. I picked it up, surprised by its weight, and took it home.

They told me that it worked as they had turned it on just before I got to their house. I carried it into my dining room and put it on the big oak buffet. I was home alone, as my wife was out working to pay for this radio and my son was with friends. I thought this is great, I'll sit down at my big old dining room table and pretend its 1942. I fooled with the dials, and turned it on. I stood looking at it and realized it had to warm up. I thought it might take a few minutes. I sat down with my back to the radio and began reading the paper. I could hear some static and then a deep voice said, "Who knows what evil lurks in the hearts of men?" I knew it was the radio program *The Shadow.* For a second I froze, the thought that I was involved in an episode of the Twilight Zone passed through my mind. I turned and checked the dial. As it turned out, I had tuned it to a radio station in Philly that played old time radio programs. What can I say? I had a great evening listing to shows to which the original owner probably listened. So here I am, 45 radios and 25,000 old radio shows later, I believe I was fated to get that '37 Silvertone.

CD as Hallucinogen
Walter Lindenbach

*The absence of hiss in a musical recording–much to be desired–
can, ironically, be difficult to accept.*

Hallucinate: to wander in one's mind.
Hallucination: a mistaken impression or idea, a delusion or error.
For example, there was a new guy at a service shop who the boss sent
to the parts distributer to get some "dB gain". The boss then phoned
the parts house and told a desk clerk that the new guy was coming
over and why. The clerk hand-labeled a bottle "dB gain", put some
cleaning fluid in it, and gave it to the greenhorn from the serviceshop.
The new guy had heard that "dB gain" was really good stuff, so he
sniffed it and became like the guy who thought he could stop smoking
by sitting on a gas tank: they both got high.

Oh, haw, haw, haw! Very funny, but that's not how the CD did
it. As the Donkey said of his nether physiognomy, "Thereon hangs a
tail." (Er, tale.)

Predisposition to Spiritual Phenomena
At 13 years of age in 1953, I saw an ad by a concern called
"Music Treasures of the World" offering a 12-inch LP containing the
Beethoven Symphony No. 5 and the Schubert No. 8 for one thin dime.
The dime was sent, the record came, the Beethoven was played, and
I said to myself, "Well, what can you expect for one dime?" But there
seemed to be something there that made me play it again, and then
again. Now, after hearing it probably more than fifty times, it is still
a joy and there are still new things not heard before.

Then, there was a Pentron tape recorder at high school that
intrigued me. Do you know what happens if you record bird-song and
play it back at half-speed, then at quarter-speed? You hear some very
interesting new sounds that are probably not expected. But in these
experiences, one always hears something else that isn't wanted, and
it could not be stopped.

Ol' Scratch in the Works

That something else is called "Ol' Scratch" or "Ol' Nick" or quite a few other names. He has a mob of malicious minions, two of whom are Hiss and Hum. Hiss is his second cousin, and they besiege record players together. Hiss also lays claim to a divine attribute: omnipresence. He is present in all tape recorders: past, present and future, world without end, amen.

I, as a true and valiant knight, have done battle with these tormentors and enemies of good music. All my life. No holds barred; all's fair in love, war and Hiss. My first tape recorder was a Philips 300. It was a good machine. Music sounded good even on it's own speaker, and better through a sound system. But of course, it had hiss; about 40 dB down from maximum average recording level. High-end roll off was -3 dB at 5 kHz, and it seemed to me it could be improved, producing a better signal-to-noise (S/N) ratio at the same time. A 24 pF capacitor across a 470 K ohm resistor that applied the audio to the recording produced a roll-off of -3 dB at 23 kHz.

Oh, goodie, so clever! But, peak levels had to be watched ever so carefully. Ever heard a beautiful soprano solo soaring up to a climax and hit the tape saturation level? It's horrible, horrible. Next the Ampex 350 and 601 recorders came into my responsibility. The first playback (PB) amplifier stage in the 350 is a 12SJ7 sharp-cut-off pentode marked on the schematic drawing "select". The first PB stage in the 601 is a 6F5 high-mu triode, which certainly should have been marked "select", too. Why is there a pentode in one and a triode in the other? Interesting, isn't it? I don't know why. Do you? It sure would be good to know.

The usual S/N ratio of the 601 was -40 dB with 10 dB headroom above VU. That's not a lot. An improvement to -45 dB S/N ratio was possible by selecting the 6F5 first PB amplifier tube. It seems that a used tube is better there than a new one! Maybe more experience helps, but 45 dB from VU recording level to the hiss is still not much for orchestra music. Ol' Scratch and his cousin Hiss were still winning.

These things happened at a radio station, where new LPs arrived in boxes of 50 or more. They were commandeered by the "engineer" (me) for inspection before they could be used on the air. This gave opportunity to select and copy to tape those works that I wanted before the records suffered from use, thereby minimizing the scratch on my tape copy. It was a handy arrangement.

One day, a violinist came to the studio for a recording session. He was an old man who really knew his violin *and* how to make it sound good on the recording. This young "engineer" set up the mic optimally for "instrument recording" by the book. After a few recording minutes, the violinist asked, very politely, if he could arrange the mic as he was accustomed. He placed the mic at shoulder-height, and when he played softly he moved the violin to about six inches of the mic. I thought, "Quick, grab the level control; he's going to go right over the top!" Not a chance: he lowered his playing volume, keeping the VU meter right in the best range and, when playing louder, he moved away from the mic, again maintaining correct level. What a delight to record such an artist! And the sound! Wow! The quiet intimate passages could move you to tears–and there was almost no hiss!

Then there was a recording session in a big old stone church one Saturday morning. The temperature was 30 degrees below zero–yes, Fahrenheit! (How do people live in temperatures like that? Like porcupines make love: very carefully.) There was a fearful hiss in the place which, it seemed, was the steam heating. The janitor was asked if the heat could be turned off to which he consented "boot on'y fer an oor, er oi naiver git daplaice warmit oop fair Soond'y marnin' kerk sairvice." So the recording session began, the music rose, and the temperature dropped. Before long, everyone had their coats on, and my fingers were well-nigh frozen to the controls, but there was no (extra) hiss.

Epiphany

In the early 80's, I was running a small company in a building with a large space above the offices. It had reasonable acoustics, and a stereo system was set up. There were four speakers: two big bass units, in the back, and two full-range units near the front. There were two amplifiers: a dual 100-watt unit for the back bass speakers, and a dual 35-watt unit for the forward full-range speakers. It sounded pretty good and after hours when everyone was gone, the volume could be turned up to a really satisfying level. The office address was Farrell Rd., so it was called the Farrell Road Concert Hall (FRCH).

During that time, a work called "Carmina Burana" by Carl Orff had captured me. It's a glorious, crazy thing. It starts with one great thunderous orchestra-and-drums note, then a couple of bars with full orchestra and chorus, followed by a moment of silence, then the chorus almost whispering.

A friend named Rick came over for a demonstration, and hear "Carmina Burana" at a fitting volume. When it started his lips moved what seemed to be short theological and biological expressions, most four-letter words. This seemed to be a satisfying symptom of surprise.

When it finished, he said, "There sure is a lot of hiss in that recording, Walt."

"Well of course there's hiss. The source is tape, and it was at a pretty high level."

"Have you ever heard of CDs?" he asked. Well sure I'd heard of them, but actually heard one? No.

Some time later Rick came over for my birthday party, and his present was a CD of–you guessed it–Carmina Burana.

"Thanks so much, Rick!" said I, "but where's the player?"

"Oh," said he, "you will have to get that."

So I did, and connected it to the amplifiers at the FRCH. With the Carmina CD in place, I pressed PLAY and there was *nothing*. I was about to look for something wrong when the full thunder of that wonderful and terrible first note filled the place. Then the chorus, then silence–no, no! Of course there was a hiss, there had to be a hiss! I've put enough tape through machines so that at five cents per foot, I could retire twice over, and there never was music reproduction at this level without hiss! Of course there was a hiss...there had to be...there must be...gradually the hallucination faded and...my god there is no hiss!

Rapture, ecstacy, victory! Hiss gone! Just a pure velvet silent background for the music. It was orgasmic! The demon was dead! Here was a complete freedom from hiss. It was what I had sought all my life. Maybe you are getting the idea that I was a bit excited. I grabbed a chair to keep from floating away–or falling over; I was dizzy!

Truth is where you find it

An experience like this can produce fanaticism. "If it's not a CD, how can you listen to it?" Now that's intolerance; not good. Consider:

I had a cousin who made "Dagwood" sandwiches. Pickles, salmon, strawberry jam, fried eggs–that was his typical five-decker version. Did he eat it? Yes, and he was in a state of ecstacy with one of those things. Did I eat one? Well, no. He offered, but I couldn't quite...Did he survive? Yes. Tummy-ache? Well, he didn't talk about after-effects. So, you see, oddity runs in my family. There's no accounting for taste.

Some people listen to classical music. Some people listen to other things they call music. Some people listen to Mahler! Many come to a state of ecstacy with such forms of music. There's no accounting for taste. Once, at a party, I described my fight against hum and hiss to a puppy (young male of genre *homo sapiens*). "Oh a bit of that is OK," he said. He seemed to imply that it was part of the ecstatic musical experience. There's no accounting for taste.

Some people don't like CDs; digital sound too crisp and cold for them. They prefer the old black vinyl records, hiss and all, and that to them is the stuff of ecstacy. There's no accounting for taste.

But if you ever have an experience of ecstasy like mine, with a Dagwood sandwich, Mahler, hiss and hum or old vinyl records, good. If one of these comes near "blowing your mind" as the CD experience did mine, good. That is very, very good, and may you have a great joy of it evermore.

My Philco 90
Rege Flaherty

I originally acquired this Philco 90 cathedral off the Internet. It was in a poor state of repair missing speaker, some tubes and many pieces of veneer. The variable capacitor was encased in a light rust and the chassis had not been worked on previously but had rested for some time in an unheated location. My first thought was to restore the cabinet and then work on the chassis, which I did. I began by first cleaning up the chassis and cabinet as best I could and then using a Dremel tool and a fine wire brush to remove rust and scale.

I spent many hours polishing up the chassis including the underside. By loosening screws holding some parts to the chassis I was able to get in those tight corners and really polish up the metal

The clean Philco 90 chassis.

chassis and associated parts. Correct tube shields were found and with the help of Chris Wells and Ted Depto, knobs and a speaker were located to complete the chassis. All capacitors (the black bakelite types) were routed out, disassembled and new modern capacitors inserted inside, finally sealing the top as original. A word of caution when using the Dremel tool. The shaft turns at around 18,000 rpm so when you're using it in tight places around the chassis, watch that the wheel does not contact wires or hidden parts since the high speed of the tool will jerk and possibly damage parts, or worse break the bit or wheel.

The front of the cabinet was missing several pieces of veneer and real walnut wood veneer was located locally. After glueing up the front rim and then stripping the sides of the cabinet, I applied a sealer, walnut stain toner (Mohawk) and finally finished the cabinet with many coats of clear semi-gloss lacquer.

First place winner.

The entire cabinet was rubbed using fine pumice, resulting in a radio that is a showpiece, and I am proud of the way it turned out. With much patience and fortitude, I was able to turn a junker into this first place winner which won at our September 2006 meet.

Golden Age Radio
Dan Piesik

Who knows what evil lurks? Is it Sam Spade, The Lone Ranger, Vic and Sade? Is it Gangbusters, Rosey Rowswell, Lorenzo Jones, or John's Other Wife?

No, it was not Ma Perkins, Don McNeil, or Fair Fat and Forty Davy Tyson. It was, of course, The Shadow. If any of you kids remember any or all of these radio shows and characters, stay tuned.

The shows and radio in general were very exciting in the 1930's and '40's, which is termed as the Golden Age. The networks were full of programming to entice listening by young and old alike, and listen we did. Our whole world rotated around the radio, the stars, music, and especially news. Every September, everyone was talking about returning shows, since personalities took three months off for vacation and summer substitute shows were the forum for new programming and potential stars. Those were the days of live broadcasts and no repeats. Jack Benny, Fred Allen, Edgar Bergen and Charlie McCarthy were the staples for comedy. Lux Radio Theater, Mercury Theater of the Air, and First Nighter held sway in drama. Of course the "Soaps" originated in radio, as the soap companies sponsored an array of late morning and afternoon dramas that persuaded mom to sit down, take a break, and buy Rinso or Oxydol soap products.

Kids on the other hand listened to serials and half-hour dramas that were sustaining, which meant that they came to each radio station on a large transcription record disk. The station inserted the advertisement at the appropriate time. These programs could be presented at a variety of times during the day or night. The Shadow, Lone Ranger, Green Hornet, and a series of detective dramas were done in this fashion. Fortunately, these transcription discs are still available and several companies copy them onto tapes and CDs. The live shows were not recorded and the only material available was from individuals who had a record lathe at home or from copies made by the legal department that were not discarded.

Let us see how many of the old shows you remember. How about Perry Mason in the afternoon, Luncheon With Lopez with Vincent

Lopez at noon, Mr. District Attorney, and The FBI in Peace and War with the theme song by Sergei Prokofiev. Then there were the comic antics of Blondie and Dagwood, Bob Burns, and People Are Funny.

H. V. Kaltenborn, Gabriel Heater, and Lowell Thomas held sway with news for the adults. Surprisingly, these guys were very honest and politically neutral. That is, most of the time.

Local radio stations included WWSW, KDKA, WCAE, WJAS, and KQV. Each station was affiliated with a major network while others were independent with a variety of program formats. Remember CBS, ABC, and Mutual?

On Saturdays, kids tuned to Starlets on Parade, which was a local studio presentation of KDKA in Pittsburgh, Let's Pretend, and Buster with Smilin' Ed McConnell was followed by Adventures of Archie Andrews. Remember Junior Miss, Billie Burke Show and Grand Central Station and Meet the Meeks from before noon to 1 p.m.? All the kids and a few adults memorized the opening and closing of the unique shows. Can you listen to the William Tell Overture and not think of the Lone Ranger?

A Hopeless Case
Paul Hyman

A friend asked that I repair a Zenith set which had sentimental ties. I agreed; that was a real mistake!

The model number was illegible but the set was distinctive enough to be identified–6S511 chassis 6A02.

The cabinet was removed and our sweetheart was exposed. The line cord was denuded of insulation in places and these had been charred. Electrolytics had been replaced at least twice. The original and replacement electrolytics were tied together. If they were open that was less of a disaster then had they been shorted.

The speaker was a good 15 years too recent; it had no field coil, which had been an integral part of the power supply, and this was replaced by a 10-watt resistor.

The voltages on the plates were much too high and sure enough the power transformer had been replaced with little attention to the original design voltages. The negative grids were also high.

Four of the six tubes were loctal and two of these were not the original types; the converter and the RF amplifier were substituted with similar but not identical type loctals.

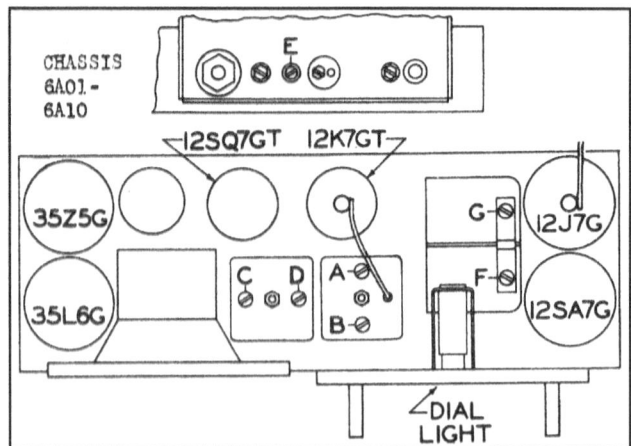

Rider's **volume 12/Zenith page 9.**

I made it clear to my friend after this examination, that resembled a postmortem, that survival was iffy. He wanted sweetheart fixed.

The cord and power capacitors were replaced and the wiring changed to restore the original relationships shown on the schematic (they had been rearranged by the hand of the past repair person). This occurred when he substituted the resistor for the field coil.

With the set fired up, I could hear general RF noise and receive the strongest station but not the weaker ones. These came in if the RF amplifier was bypassed when a long wire was connected to the signal grid of the converter. I judged it was time to get the correct tubes into the converter and RF sockets. The RF amplifier came right out but the converter resisted. Prying moved the tube and a mighty pull removed it. Unfortunately parts of the socket never left the tube, the socket was disintegrated and sweetheart had expired there on the table.

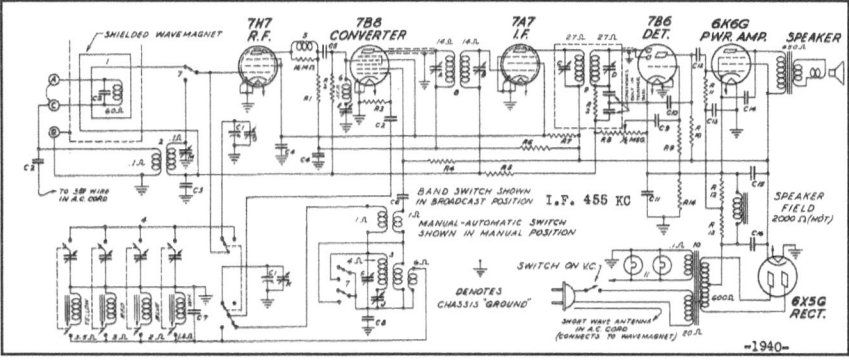

Rider's **volume 12/Zenith page 13.**

Small Receiving Antennas
Paul Hyman

L et us discuss small receiving antennas.
   I was faced with repair of an attractive wooden Philco radio, date and model unknown, but of distinctive design.

Using Ramirez's site *philcoradio.com* and methodically following instinct, I found the set, a PT-44. It was introduced in June 1941, cost $11.75 and 11,303 were made.

This information gave me the circuit diagram via *Rider's* on *nostalgiaair.com*. It was found in *Rider's* volume 12/Philco page 45.

I had the wooden cabinet, wooden back and chassis. The circuit diagram made it clear there was a loop antenna, but it was not found on the radio.

The filter capacitors were changed forthwith, as were suspiciously worn capacitors, the 47 pilot bulb, and the line cord.

It was only then that I applied 115 AC volts via my fused isolation transformer.

No tubes or pilot light; the filament chain was open. Rather than go to the tedium of tube testers, filaments were probed with an ohmmeter after scraping the pins of the loctals and single octal free of accumulated debris.

The 50L6 filament was the fault. When it was replaced, the set lit and the glow was heartwarming but not the noise that no signal could be coaxed out of.

My test equipment includes a scope, home brew battery powered 455 kHz modulated signal generator, and a one kHz audio generator. After I applied these tools, the audio and IF sections where functioning and the scope demonstrated local oscillator function.

The tuned circuit of the signal grid of the 7A8 was without an inductor (the loop antenna) hence no signal. The loop was necessary, it was both antenna and the inductor for the tuned circuit.

I proved this by attaching a Meissner antenna coil to the wires left dangling from the loop's attachments on the chassis. I was able to hear only the strongest of stations; the coil was not the exact inductance and its capture area was much less than the loop, but it worked. My surmise was right.

Small loops are antennas whose conductor length is defined as less than 0.1 - 0.085 of the wavelength they are designed to receive. The broadcast band covers 540 to 1600 kHz, about 562 to 180 meters wavelength. Small loops are sensitive to the magnetic field of the electromagnetic radiation, large loops (as in the order of one wave length) and dipoles are sensitive to the electric field.

Voltage induced in the loop by a field of E strength is described by the equation

$$V = \frac{2\pi\, ANE\cos\theta}{w}$$

When a capacitor is place parallel to the loop and the inductive reactance is matched by the capacity of reactance, resonance, the induced voltage is multiplied by the Q of the tuned circuit.

Properly designed loops can attain a Q of 100. As seen additional factors affecting induced voltage are A (area of the loop)   and N (number of turns).

The next parameter to be determined is the loop inductance. Fully meshed the capacitor tunes the signal grid of the 7A8 to 540 kHz, the low end of the BC band. That capacity will dictate what inductance was needed.

Here I use the ALMOST ALL DIGITAL ELECTRONICS L/C meter IIB. This device accurately measures inductance and capacitance. It read the total capacity at 540 kHz across the loop as 564 pF. The variable capacitor in this radio was indeed larger than the typical 365 pF units.

The formula for resonance is

$$F = \frac{1}{2\pi\sqrt{LC}}$$

At 540 kHz the inductance needed to resonate a 564 pF capacitor is 154 microhenrys.

There are two varieties of loops used in "modern" (after 1939) radios. The early ones used open planar loops. Later smaller loops used ferrite core small solenoids. This enhanced permeability and reduced size became standard. I chose, despite its temporal incongruity, the ferrite coil to achieve as much signal in less space.

The reluctance of the ferrite is less than the surrounding air; this causes the signal field to pass through the loop rather than pass by it. Magnification of loop area, and thereby intercepted signal, results. A small ferrite coil is thus equivalent to a much larger air core loop (about 18 times larger).

Inductance as well as Q are magnified if the coil is centered on the rod. Output voltage is directly proportional to loop area, turns and Q rod permeability, and inversely proportional to wavelength.

Using the LC meter I empirically added turns at the rod's center for a reading close to 150 microhenrys.

IT WORKED !

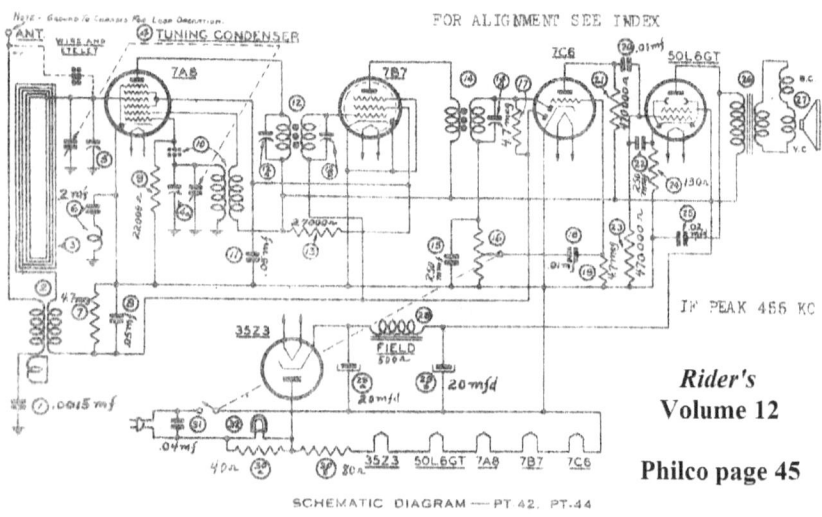

SCHEMATIC DIAGRAM — PT 42, PT-44

Rider's
Volume 12

Philco page 45

RCA Resurrected
Richard Brewster

Way back in the forties my brothers and I used to sit on the floor in front of a large console radio. The RCA Model 9K2 was given to our family by my aunt who purchased it new in 1936. I have had such fond memories of that set! But unfortunately, at that time, I was just becoming interested in electronics and had great fun 'taking it to bits,' as the British say.

RCA Model 9K2. Photos by the author.

Some years ago when we were living in the south hills of Pittsburgh, I decided that I would like to find another. I advertised in a national publication and, sure enough, one turned up. Not in Wisconsin, not Texas but in the east end of Pittsburgh!

And then the owner, another collector, was willing to trade for it. So it didn't even cost me anything! And it was in really nice condition to boot.

I did a 'quick and dirty' restoration just to get it going and really enjoyed it for a time. Then last month, I decided to do it right. First of all, all paper capacitors and electrolytics were replaced. No tubes were bad. I had previously replaced the 'eye' tube with a NOS one.

Back in 1936, power was 110 volts, not the 120-125 that many systems now supply. So I installed a 12-volt filament transformer, wired as an autotransformer to reduce the incoming voltage to around 110. Then, I wired a thermistor in series with the power transformer to eliminate the surge when the set is first turned on. And I've replaced the cloth power cord. Does that set ever work well. No alignment was done but the dial calibrations are right on, and now I can keep it on for hours.

Close-up of 9K2 dial.

My FADA 930 TV
Tom Mizikar

I had a Fada 930 TV for the November 1st, 2008 clinic. I brought it in because I had previously (about two years ago) replaced all the capacitors, the mica caps in the IF section, a ton of off-value resistors and a few other things ... like the CRT and a shorted control.

The set is a bit unusual in that it is a Fada and has a blond or light oak finish instead of the more common mahogany. It is an RCA 620 with a 12-inch picture and I'm sure it was considered to be a "better" TV than the smaller RCA version (even though it's the same thing). I'm sure that the two-inch larger screen was impressive in its era.

After all that work the set still didn't present a correct image. It has a raster and synchronizes the picture but the image is stretched out on the left and folding over on the right. While I am developing a basic understanding of circuits and radios, I'm really in over my head on something so complicated as a TV. Fortunately for me, I viewed working on the TV as an opportunity to advance my knowledge instead of a fear-inducing effort. And, fortunately too, there are people who know what they're doing and are willing to help me.

Tim Tress had been providing me with some help in an online forum (his Antique Radios) and was helpful by offering to take a look at it in person at the PARS meeting. Tim fired it up and, without test equipment, here's what he found:

> *the picture is stretched on the left side, folded over on the right side, and the width and horizontal linearity coils have no effect when they are adjusted. I would carefully re-check the wiring and all components around the 5V4 and the linearity coil, making sure that nothing got misconnected when you replaced the capacitors. I turned the width coil slug most of the way out with no effect, so screw it back in until we get the rest of the circuit working.*

Tim also suggested that I try a new 1B3 and check the resistors and capacitors around the drive control. I would save checking the capacitor inside the yoke for last, due to the trouble of removing the ion trap and focus coil. He also added, "the focus control brings a picture into focus at one extreme end of its rotation; the resistors around the focus coil and the focus coil itself should be checked."

As I was keeping notes and getting input from Tom Dixon, Karl Laurin, Lou and others, Karl offered to take the set home and see if he could get it straightened out for me.

My hope is that I can get the TV working well enough that I can enjoy using it a bit until the analog signals go off the air and also use this set on the night the analog signals go dark…watch the end of the analog era on a set from the very beginning of the era.

Many thanks go out to everyone who took a look at it, and especially Karl and Tim for their help and expertise. If Karl is successful, look for the TV at an upcoming PARS meet.

'Hands-on' Troubleshooting
Joe Patrick

Troubleshooting old radios can be as easy, or as difficult, as you make it.

While most radio repair technicians have and use a multitude of test equipment, I have known more than a few 'old timers' who could quickly troubleshoot most radio problems with nothing more than a VTVM (vacuum tube volt meter), tube tester, soldering iron, a few screwdrivers and hand tools, and their fingers, eyes, ears and nose. Many times, they did not even need or use a schematic!

My old friend John was such a person and he taught me much in my early teenage years about radios, servicing and electronics. He taught by example and proved to me many times that simplicity was often the best, quickest and easiest way to analyze and repair old radios and electronic equipment. BUT, he could do this only because he had a complete understanding of tubes and vintage electronic circuits and radio designs.

John said that there are primarily four 'trouble spots' that cause 90% of most radio failures.

1) Tubes 3) Capacitors
2) Power supplies 4) Resistors

Occasionally other problems would arise in coils, I. F. transformers, speakers, solder connections, etc., but to first concentrate on the above four areas to solve most problems.

In servicing a radio, John said that all tubes are to be tested first and any defective ones replaced. If this step did not cure all problems and return the radio to 'factory-spec' operation he would move on to the next steps.

Invariably, I observed and was taught by John, to check the power supply next. He would always power-up his radios using a 'dim-bulb' test adapter–usually switching between 40, 60 or 100 watt bulbs as required by the current drain (wattage consumed) of the radio under test. From his many years of experience, he could tell a lot about the radio's operational condition just by watching the glow of the dim-bulb's' light bulb and by how bright or dim it was. Power supply shorts would quickly cause the bulb to illuminate 'full on'. Normal radio operation would cause a gradual rise in illumination and a much softer illumination–as long as the correct bulb was used to match the normal current demand of the radio under test. No illumination indicated that the radio was dead or not powering on.

If John determined that the radio had a short, he would usually check the selenium rectifier (if used) and large, electrolytic capacitor(s) with his ohmmeter. If no shorts or high leakage were determined he would power the radio back on and then feel–with his fingers–the larger power supply resistors to see which one was getting excessively warm or hot. Once he found it, he knew it was drawing excess current and would then trace its circuit to determine the cause of the problem. Sometimes, he would use his nose to 'sniff-out' problems, or just simply look for burned or discolored resistors, 'wax-melted' capacitors, transformers and power supply chokes and other components.

If John determined that a set was not shorted by watching the 'dim-bulb' he would usually begin his troubleshooting by taking some power supply AC and DC voltage 'readings' with his old Heathkit VTVM. By doing so, he could tell much. Such as if the AC circuit and the 'on /off' power switch and power transformer were working; or, if the rectifier (tube or selenium) was operational. Open power resisters could also be determined as defective and then, with the power disconnected, 'ohmed-out' to see if in fact they were open or had markedly increased in all ohmic value or had 'opened up'. Sometimes a bad solder joint would be 'eye-balled' and found. Once John has was certain that the power supply was operational, he would move on to the next phase of his troubleshooting method....which was the signal path.

If no audio was heard through the loudspeaker, John would use his finger to touch the center tap of the volume control. If hum was heard through the loudspeaker, he knew that the audio circuit was most likely OK from the volume control to the speaker and thus would concentrate his troubleshooting to the radio's 'front-end' stages. As before, he would use his fingertip or a small screwdriver–with the metal blade held between his fingertips–to touch the grid connection pin or grid cap of each 'front-end' tube to determine if any hum or noise could be heard through the speaker. This technique would provide some insight as to which radio stage was inoperative. Once the defective stage was known, he could then use his VTVM or other equipment to find and determine the cause. John would also often use a 'neat' little signal injector pen at times to assist in signal-tracing a problem. I think his pen was either a Don Bosco 'Mosquito' or an Eico PSI-1. I have a 'mosquito' myself and use it quite often. The 'cool' thing about these injector pens is that they produce many harmonic frequencies that can be used to inject both RF and AF signals–thus enabling circuit tracing through the radio's RF, IF and audio stages. At times, he also used that old Heathkit signal tracer with much success.

A few times, I observed John using another radio–in good operational condition–to determine if a defective radio's oscillator circuit was working. He would adjust a good radio's tuning to a clear spot near the middle of the dial and then turn the defective radio's tuning knob while listening for a heterodyne (whistle) or carrier signal on the good radio. This 'trick' would let him know if the defective set's oscillator was working or not. John had many tricks like this, all learned from a lifetime of reading, radio servicing and experimenting. He was a real 'hands-on' type of guy. From touching grids to using a grounded capacitor to intentionally 'kill' signals–to determine the source of noise-related problems–John made servicing radios look simple and he was extremely good at it!

I remember one time he was repairing (now called restoring) a Hallicrafters ham-band receiver, of which the band selector switch had caught on fire. The switch was 'TOAST'! John told me that he was going to fix it ... I thought to myself ... Sure !

As I recall, the band switch was a rotary, multi-pole type, with at least four or five sections of Phenolic wafers and many solder-attached wires. The contact pins were the typical finger-style wafer-switch pins that are held in place by very small rivets. Believe it or not, someone used to make a kit for repairing these wafer switches, complete with the rivets, wafers and individual 'finger' contacts. John happened to have one of these repair kits and set about to rebuild the burnt-out sections of the switch. Over the course of the next few weeks, John completed the repair as I watched in awe! The radio worked like new, too!

John knew when it was time to bring out the 'big guns' and he had an arsenal of test equipment such as an oscilloscope, AF and RF signal generators, signal tracer, condenser (capacitor) checker, decade boxes and much more. While used at times, he was good enough at 'hands-on' troubleshooting to keep test equipment use to a bare minimum.

My friend, mentor and fellow ham radio operator John passed away in 1994. He was quite a guy! He taught me much and fanned the flame of my early radio interest. I can't help but think of the knowledge and information about radios, servicing and radio history that is lost as each 'old timer' passes on, taking their first-hand, living history of radio experience, knowledge and skills with them. I know for me, some of that knowledge continues with each 'new' old radio that I restore. And, I am sure too, that each time I accidentally get the be jabbers 'zapped' out of me while working on a 'live' set, John is looking down ... laughing, while saying ... "Remember Joe ... keep that free hand in your pocket!"

TV booster by Knight, a division of Allied Radio.

A Television Demonstration in 1928
Raymond M. Bell, Dickinson '28

On February 29, 1928 at 6:45 PM, I gave to a crowded house a demonstration of closed circuit television. It was in the lecture room of the Tome Scientific Building on the campus of Dickinson College, Carlisle, Pennsylvania. It was done under the guidance of my physics professor, Fred Mohler, who had studied under Rowland at Johns Hopkins University in Maryland.

I had read in the newspaper of the television broadcast on April 7, 1927 by W3XN Whippany, New Jersey on 1570 kHz using 50 line - 18 pps pictures. A junior in college, I asked Dr. Mohler if we could do anything in television. He said we did not have the facilities by radio, but we could try using a closed circuit.

Dr. Mohler indicated there would be three problems in this endeavor:

1. Mechanical–making rotating disks for each end of a shaft driven by a motor.

2. Electrical–amplifying a signal from a photoelectric cell, passing it on to a neon tube. (Light on a photoelectric cell produced electricity, and conversely electricity on a neon tube produced light).

3. Optical–focusing the light from the filament of a 200-watt tungsten lamp on a rotating disk and at the other end viewing the light from the neon tube through the rotating disk; lenses were needed. So the first step was to learn about photoelectric cells. This I started on May 2, 1927. By May 9th, I had learned how to amplify the current put out by the pe cell and had learned about a neon tube. In the fall I resumed work and recorded in my log of November 21, 1927: Worked on one stage of amplification on a pe cell. I worked with a neon tube in circuit to produce an effect on it when light shines on a pe cell.

My next task was to take two circular metal desks and bore 24 1-mm holes, staggered. When rotating the sequential holes formed a picture about ½ inch by ½ inch.

One looked through the rotating disk at the neon tube and saw the image of the tungsten lamp filament. With both disks on the same shaft synchronization was no problem.

The light from the filament shown through the tiny holes of the rotating disk to the photoelectric cell. This impulse was amplified and sent to the neon tube. The tube glowed and was viewed through the tiny holes of the other disk.

By February 13, 1928, I had a system working with 24 lines at 18 pps. I made a public demonstration on February 29, as stated above, and on April 14, a private demonstration to my cousin Clarence A. Bell. He and I had worked with radio from 1922.

On July 2, 1928 television became available to the public when W3XK Jenkins Laboratories, in Washington, DC began sending television pictures on 1605 kHz and 6420 kHz at 48 lines, 15 pictures per second. On July 10, 1930 Clarence and I at Lewistown, Pennsylvania began receiving pictures from W3XK.We used a kit made available by Dr. C. Francis Jenkins, an American television pioneer.

Super-Sonic booster manufactured in New City. Leatherette covered case.

The Breting 12 Receiver
Joe Patrick

A classic 1935 "scientifically correct" DX radio.

L ittle did I realize the journey I was about to take from simply
reading an online craigslist ad –it was an estate-sale advertisement
for liquidating the contents of a once very active Pittsburgh-area
electronic repair shop.

I called my friend Rafael (RAF') to ask if he would like to go on
Saturday to check it out. "Of course!" he said. So that weekend found
us scrounging through the contents of this once significant repair shop
seeing what "trophies" we could score and take home.

We each quickly found and purchased a few items; continuing to
dig our way through piles, boxes, cabinets and shelves full of old
electronic parts and equipment of all types. This shop had been in
business for many years and all kinds of vintage merchandise was just
waiting to be claimed for the right price. Of course nothing was
marked...so prices had to be negotiated on everything. But bargains
were to be found, as my car was full by the time we left an hour or so
later.

I managed to buy a vintage Fidelitone phonograph needle display
stand complete with many old needles. One of my "prize" purchases
turned out to be a box full of various miniature pilot lamps–of the type
used in vintage radios for dial lights and such. There were hundreds
of them...all new-in-box–for just $10.00!

During this "shopping spree" I spied a couple of vintage ham
radios. One was a Harvey Wells BandMaster transmitter and the other
a Hallicrafters SX-25 receiver. On the floor, with a bunch of stuff
piled on top of it, was an old, dirty Breting radio receiver. *A Breting
what?* I had never heard of this brand before. It looked to be in pretty
sad shape and it was easy to see that a previous owner had *heavily*
modified it. This thing was calling out to me... **SAVE ME**! I knew
that I had to have it... So I finally asked ... how much?

To my *total* surprise the guy selling the stuff said..."How about
fifteen dollars? It's pretty hacked up; you can use it for parts." Did I

The Breting 12 receiver as found and purchased.

hear him right? *Fifteen dollars!* Without even cracking a smile I said, "Sure, that sounds good to me". It wasn't until I returned home later that I fully realize what I had done…or gotten myself into.

Breting Radio was started by Paul J. Breting. The chief engineer was Ray Gudie who also designed the Patterson receivers. The Breting model 12 was Gudie's first major design for Breting and was proclaimed to be a "scientifically correct" DX radio. Breting built receivers from 1935 to 1940. With the exception of the Breting 6, most were high-end receivers with high tube counts, chrome chassis, crystal filters and many other features not available on Hallicrafters and National radios of the same vintage. All the Breting chassis were manufactured in the Gillfillan plants–the model 12 at the 17 Venice Boulevard, Los Angeles, California location.

My Breting 12 purchase is a big, heavy, solidly-constructed 50+-pound receiver with an 18-gauge chrome-plated chassis and transformers, and highly-polished aluminum tube shields–similar in design to Scott's and other high-quality receivers of the time. The Breting 12 offered 550 Kc to 32,000 Kc 5-band reception using 12 tubes, dual panel meters and a 432 Kc crystal I.F. filter. The 18 watt audio section uses a pair of 42's (push-pull) driven by a 42. It requires an external 12" electro-dynamic speaker. Built in 1935-36, it listed for

$155 complete with black crinkle finish metal cabinet, tubes, external speaker and crystal I.F. filter. Amateurs and experimenters could receive a 40% discount off the list price, so in actuality, it sold for $93.00 F.O.B. Los Angeles.

This all sounds good on paper, *however*, my Breting 12 was *heavily* modified by someone and nowhere near to original condition. Whoever previously owned it… It looked as if they owned stock in a toggle switch company, as they added 11 toggle switches to the front panel! There were also numerous other circuit and mechanical modifications and additions. This was going to be one "tough cookie" for me to restore back to "original" condition! *But*, it is just the type of challenge I love and thrive on !

Where To Begin?

Before I begin any restoration, I go online to find as many photos, schematics and other information as possible. This helps me *immensely* in gaining knowledge about the particular set and in formulating my plan of attack.

Breting chassis topside.

All of my restoration projects begin with lots of mental imagery. I mentally break down the project into small steps and carefully think about how I will begin and accomplish each one. I can't stress enough how important it is to do this. It is *very* important
not to try and think about all that needs to be done at once or you may feel totally overwhelmed by the scope of the project. Just plan and do each small step one by one and the project will eventually take care of itself.

A Few Problems...And Solutions

After studying photos of other Breting 12's, a few things became immediately noticeable to me. First was the addition of a large filter choke to the underside of the chassis. In tracing its wiring, I determined that it was used to replace the field coil of the electro-dynamic speaker, so that a standard permanent-magnet speaker could be used. This choke also had a 1500 ohm wire-wound power resistor in series with it to simulate the winding resistance of the field coil. Both of these items were removed as I want everything back to the way the receiver was originally designed.

Breting chassis underneath.

I also noticed some expected repairs and additions, such as new power supply electrolytic filter capacitors bypassing the old metal-can chassis-mount types that had become defective with age... and a few other capacitor and resistor replacements.

One modification had me puzzled for a while. The Breting 12's I.F. frequency is a somewhat odd 432 Kc, and as expected, that is also the frequency of the crystal filter. But in my set, the crystal filter frequency is marked 456 Kc. Why?

I thought long and hard about this until one day I was looking up the specs on the Heathkit QF-1 Q multiplier that the previous owner had connected (hand wired) into this receiver. I discovered that it only works with I.F. frequencies of between 450 Kc-460 Kc. No way would it work at the Breting I.F. frequency of 432 Kc. So, the owner of the receiver changed the I.F. frequency and re-tuned it to operate at 456 Kc instead of the proper 432 Kc. I am now searching for the proper 432Kc crystal filter but so far no luck. Once obtained, the entire I.F. will need to be re-aligned back to its proper, as designed, 432 Kc operating frequency. (Good news! A 432 Kc crystal filter has been found and purchased).

One of the biggest obstacles to overcome was what to do with the mutilated front panel–and all of its eleven added toggle switch holes. Finding a replacement panel from another model 12 was going to be just about impossible. So that left the option of either making a new panel or somehow repairing the one on the receiver. I chose to repair the one I had.

After thinking about my options...TIG, MIG, body putty, metal plugs, epoxy, etc., I decided to try leading-in the holes with plumbers solder, flux and a Propane torch. First, I countersunk each hole on both sides using a countersink bit and my drill press. This provided more "edge" for the solder to stick to and enabled better "feathering" into the metal panel when sanded. Before soldering each hole, I backed it up with a concrete paver brick and a piece of sheet aluminum. This helped to keep the solder flat and level on the back side of the panel and the aluminum sheet made it nonstick. I intentionally "mounded-up" the solder on the front side of the hole so that it could be sanded flush using a palm sander and then by hand

The front panel with its unwanted holes filled in.

sanding with 400-grit wet sandpaper. Auto-body "spot putty" was used to fill slight imperfections and then wet-sanded smooth. This method of hole-filling worked extremely well and I am very pleased with the result.

Another modification made by the previous owner was to flip the B.F.O. control so that it came through the front panel instead of underneath the bottom of the receiver as was designed. I could understand why the owner made this modification as it is much easier to adjust up front, but again, that's not the way Breting designed it to be. In order to reverse the modification, I had to make a "new" mounting bracket for the B.F.O. variable capacitor to mount to. I had no dimensions to work from...only photographs. So using them, I was able to determine the bracket's dimensions from transposing dimensions from the variable B.F.O. capacitor, which I had, to the scaled photos of the bracket.

One thing I did not want to do was to use new, shiny metal for the bracket's construction. So I searched through my junk bin and found an old radio chassis with the correct metal thickness and aged cadmium plating. Once cut out and fabricated, it looks to be original to the Breting and does not stand out as being "new-looking" or "not proper" to the radio.

Conclusion

The restoration continues! I still have much more to do, but as I stated previously, it's one small step at a time until the project is totally completed.

My next task will be re-stuffing the tubular and electrolytic capacitors and "dog bone" resistors. I have a unique way of rebuilding "dog bones" that I will share later. Of course there are many other things to do, but eventually this project will be complete and I can then look back on my efforts while enjoying the fruits of my labor.

If anything, I hope this Breting 12 restoration project and article serves to motivate others to attempt more difficult restorations. Vintage radios are truly worth the effort in saving them, for others to enjoy and for our own collections, satisfaction and listening and viewing pleasure. Please make every attempt to save and pass them on to the next generation of radio collectors!

Breting R-meter.

Plaskon Radios
Chris Wells

Most of us are familiar with Bakelite radios but not so sure about Plaskon. If you have an ivory-colored plastic radio from the thirties that is not painted it is most likely Plaskon. It also comes in bright colors or even black and can be confused with Catalin but it has a more opaque pastel-like quality to it. This plastic is a thermoset that is formed like Bakelite but with less heat and pressure. Thermoset means that the molded part will not re-melt under heat but can burn. If you have a radio case that you think is Plaskon look close at the casting and you may see ripples that are unique to Plaskon. With the lighter colors light shines through Plaskon easily. The light from the tubes and dial light give the radio a distinctive glow.

Bakelite was patented by Leo Baekeland in 1909 and is found in most early battery sets of the twenties. Catalin is a translucent form of Bakelite without the fiber and was started around 1928 but was not used in radios until the late thirties. Plaskon was first developed for the Toledo Scale Company in 1931 to provide lightweight housings to their products and quickly spread into the use of radio housings. Plaskon is a urea formaldehyde cellulose (wood flour) mixture. In comparison Bakelite is a phenol and formaldehyde fiber (often asbestosis or wood flour) mixture. Beetle plastic is a form of Plaskon molded with color additives in a swirled fashion. During the molding process the chemicals polymerize into linked chains reinforced by the fiber.

Plaskon is more fragile than Bakelite and often has stress cracks in it. Collectors tolerate these cracks but the smaller and less noticeable the better. Manufacturers often made the same radio housing out of Bakelite, painted Bakelite and, for a smaller volume, Plaskon and even Beetle plastics. A fun quest is to collect the various plastic forms of a favorite radio. The rarity of the Plaskon and Beetle forms and the colored Plaskon over ivory makes these versions more valuable from a collecting point of view. From a distance the white painted Bakelite and the Plaskon look similar but up close the translucent quality of Plaskon is unique.

Many manufacturers and models of radios stand out in using Plaskon. The Air King mini tombstone that looks like a deco building is a good example. The Detrola Pee-wee came in both Plaskon and Beetle. Emerson made many of their radios in both Bakelite and Plaskon; the 108 U5A is a good example. The frog-eyed Setchell Carlson radios came in a variety of ivory and bright Plaskon colors as well as Bakelite. The machine aged GE and Sonora push buttons came in Bakelite, painted Bakelite, Plaskon and Beetle. So now that you know more, pick out those Plaskon radios in your collection to bring to the meet.

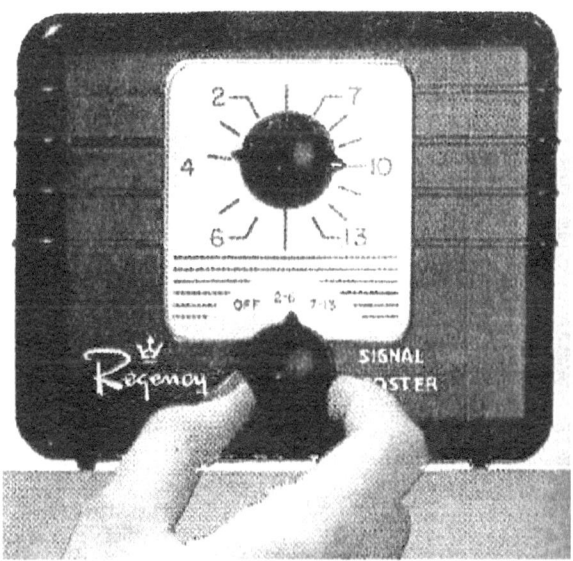

TV booster made by the same company that marketed the famous Regency TR-1 transistor radio.

Adventures in TV Land
Ted Depto, P.E.

Lately, I've been working on a 7-inch Philco black and white TV from 1948. Weighing in at 60 pounds, this little critter (48-700) is not only heavy, but taxes the parts bin for original replacement parts to get the set up and running. Speaking of heavy, I often wonder how back in the early TV days I ever carried those 23-inch color TVs up stairs and into apartments when I was active.

For one good thing though, it uses electrostatic deflection with a 7JP4 pix tube which makes it much easier to diagnose scan problems than those sets with flybacks and yokes to contend with. After checking the usual voltages and replacing a few filter caps, I brought the voltage up with a Variac. I was able to get a raster on the screen but alas, no vertical deflection on the tube. Most voltages appeared within tolerance and I began to suspect one of the timing capacitors in the oscillator stages. It uses a 6SL7 as a vertical oscillator and two 7C5 Loctal's as pushpull vertical output. Pulling out the scope, I determined the vertical oscillator was running the proper sawtooth waveform but the signal quickly diminished after passing through an .05 capacitor into the first grid of the 7C5 output. (I am not in the habit of indiscriminately changing all wax and Black Beauty capacitors in a TV before I get a usable picture. My reasoning is that you can induce more problems and make a simple repair much more difficult to solve, although most assuredly 57-year-old capacitors will fail eventually and will need to be changed at some point.)

Disconnecting one end of the .05 in question and attaching a new Orange Drop brought a full raster, but now the screen brightness was down (pix tube weak?). No, it checked good on a B&K 400 picture tube checker so I eliminated that part of the equation.

Now I am into the video control circuits and video amplifier. With the schematic in hand and a few resistance checks I found a shorted 1N34 dc restorer diode. Replacing this made a world of difference as the picture on the screen now had adequate brightness and good contrast.

After this, as I was quietly musing on my success, my mind drifted back to another time, one in which I operated a TV business, and some of the challenges that each new day provided. Some cases in point . . . .

A Hot Conversation

A service call came in on a split chassis Philco about 1954 vintage. The owner said she had a small picture in the center of the screen and it had been getting smaller over the past several weeks. When I arrived at the apartment house in which the suspect TV was located I quickly recognized the set as one in which the selenium rectifiers were a cause of concern. This was before the high current silicon diodes were available to the servicing industry. After turning off the set, I felt the seleniums and they were as I suspected, very hot. I told the lady I would remove the two chassis to the shop where I could work more freely on the set. With a jig I had made up of a 17-inch picture tube with a universal yoke and with adapter cables, I could check virtually any TV without having to lug the entire set to the shop. The picture with the test tube appeared small so I measured the B+ voltages in the power supply which was the next order of the day. The voltages appeared normal. But having had previous experience with short-lived selenium rectifiers, I decided to change them anyway. I did, and the picture widened out about an inch on either side, so I assumed it was ok. I let the set cook for a few hours, cleaned the tuner and controls, and buttoned it up for return to the owner.

I called the owner and proceeded to install the set in its cabinet. Lo and behold when the screen lit up, it was the same as when I saw it the first time–small, about 7 inches in the center of the screen. And it also appeared slightly snowy on a local channel. When I went to pull the power cord from the wall outlet, the plate covering the receptacle was too hot to touch. I measured the voltage going in the TV and found it to be 80 volts instead of the normal 110 volts ac. By taking power with an extension cord from an

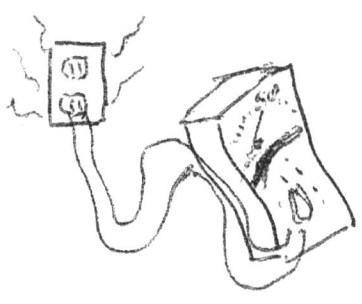

outlet in another room, I found the picture immediately blossomed and the high voltage crackled as the set came on again. Needless to say I insisted she call the apartment manager immediately and notify an electrician for corrective action. Following up after a few days, I was informed that the power had been restored to normal after changing the receptacle.

A Zap in Time Saves Hide (Mine)

The owner of this huge 27-inch Majestic floor model TV was also the wife of a renowned bolt and nut factory owner. They were pillars of the community and had a lavish home in a better section of the city. When the lady of the house called and expressed disappointment that she could not watch her favorite programs, I quickly dispatched myself to save this damsel in distress.

On arriving, I found her large dog was more than anxious to play around. It would not let me near the set after I set up the mirror and, further, the lady exclaimed "He's only playful, and really hasn't bit anyone lately." But this playfulness was getting my goat since I could not reach into the cabinet without a nip and a shove from the dog.

The lady left the room and I proceeded to work on the set. No horizontal sync. I replaced the 6SN7 sync separator and, in the mirror, the picture was stable and bright. When I began to place the back on the TV, the dog lunged at me, knocking me over. I decided it was time to act. I retrieved a test lead from the tube caddy, clipped it to the high voltage cup at the kinescope and plugged in the cheater cord. When the dog came again, I gently touched his nose with the business end of the lead. A yelp ensued, the dog reversed his steps, left the room and I never saw him again. After I presented the bill and the lady paid me, she thanked me for my work and I left feeling just fine. It was a beautiful Tuesday afternoon.

Christmas in June

Many times you hear of quirky solutions to some old TV reception problems. But this one I think is unique A channel 10 Yagi was installed in the attic of this retired Penelec employee. I took the service call, and replaced the 17JP4 picture tube with a Sylvania 85

Superbrite on his Sparton console that afternoon. After setting up the controls, I noticed a slight ghost on live programming, and with the new tube it was distracting. I mentioned this to the owner and he indicated he would show me where the antenna was and I might be able to adjust it for a better picture.

I removed myself to the attic and what did I see? He had tin foil Christmas tree icicles hanging on the tines of the antenna. I knew better than to remove them, so just a small nudge on the antenna and the picture was perfect.

No cable, no dish, no wide-band TV for sure, but this was then. The owner told me that after several days of his trying to get a clear picture, he finally discovered the tin foil scheme.

A loud commercial on the little Philco finally snapped me back from my day dreaming to the present, and looking at the clock on the wall, I shut off the Philco, checked the security system, turned off the lights and left the shop as I had so many, many times before.

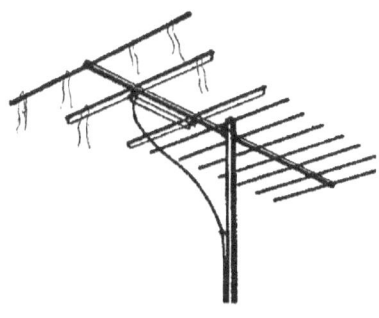

Can You Hear Me Now?
Ted Depto, P.E.

To Our Readers: The contents of this article are true and revealing even though they seem to be out of a fiction novel. Only the names of the guilty have been changed to prevent them from being totally embarrassed. I don't know if you ever had unfortunate results as I did with this seemingly innocuous experience, but it certainly makes one wonder if today you can ever depend on others to ship something as fragile as a console radio.

If I am screaming in your ear, I apologize but the incident(s) that I describe in the following should not happen to anyone, especially a radio collector. Read on . . . .

It was the last day of the month and, as always, I like to surf the internet since as your editor I generally get some ideas for future articles and occasionally buy a radio. Sometimes on the various websites, I check the wanted ads, for sale items and upcoming auctions as a matter of course.

One ad in the middle of the for sale list piqued my interest, and I decided to call the seller to find out more about it–a large, rarely seen console radio which he indicated he would not ship. The item was located in Missouri, and with me in Pennsylvania, it didn't take much figuring to know a potential problem already existed if I decided to purchase the radio.

A Good Buy

Well, against my better instincts, I purchased it with the understanding the vendor would take it to a packaging and shipping place so that it could be sent by freight. The item's dimensions exceeded both USPS guidelines and Fedex, as well as the weight, so freight was the only alternative other than driving to Missouri, which I ruled out.

I sent the money in the form of a cashier's check to insure he got it, and asked if he would deliver the radio to a packaging facility near his home. He agreed, saying "It's no big deal to me; I'll be happy to drop it off at your discretion," so after about seven telephone calls and two hours to various companies that do this kind of work, I settled on a UPS packaging store located on Taylor Road near the town of Wildwood, Missouri. I talked to a Brydan, an employee and manager of the store, who said he would, upon receipt of the radio, have it properly boxed so that a freight company could pick it up for transport to Pennsylvania . . . .

I asked the packaging place (UPS) to remove the large chassis and speaker, box it and ship it by UPS to me and I would pay extra for the trouble they had to go through. I paid for the work with a major credit card. In the meantime I must have made ten calls (Note that the telephone calls are becoming expensive) to find a freight company located in the immediate area who would be willing to transfer my radio to a long distance hauler. I told Brydan that a transfer company known as K & R Express would pick it up at the address he provided to me. I then called K & R (located in Chicago) and told the dispatcher where the packaged cabinet was to be picked up. The next day I called K & R to find out if they were able to pick up the radio cabinet, and the dispatcher told me his driver went to the address given to me by Brydan, and it was a vacant storefront!

I then called UPS packaging again and talked to a Dave who told me they moved several weeks ago to a new spot and inadvertently didn't give me the new pickup address. At this time I also told him what labels to place on the packages including ones I made up and forwarded to him I call K & R again to give the new address, and the dispatcher said, "No problem, it would be picked up today."

So Far So Good!

Next day I called the UPS packaging store in Missouri and the manager told me the cabinet had been picked up by K & R the previous afternoon, and so I thought all was well. About a week later, I received a box delivered by UPS at my front door. It had the return address of the UPS packaging store in Missouri, so I quickly opened it. To my chagrin, when I sifted through the peanuts packaging material, the radio chassis that I sought to have delivered safely by sending it separately, was just thrown in a cardboard box surrounded with peanut packaging, was severely damaged and, to make matters worse, no speaker was with it.

The radio must have taken a pretty good hit to dent the corner of the heavy steel chassis, but there it was, damaged. I called UPS to alert them, and they instructed me to save the box, packaging and all related materials, and someone would be out to look at the package. Two months later and half a dozen calls to UPS, I am still waiting.

There Is More

This is not the end of the story. Rather, at this point, it starts to become bizarre. I heard nothing about the large console cabinet for several days after I received the chassis, so I decided to call K & R Express. Guess what? The company went out of business that Friday afternoon, just two days after picking up my cabinet. The dispatcher told me he was just trying to appease customers, and did not know where the radio cabinet was located or how long the phones would be running at K & R. Perhaps it was now in Chicago, maybe in Pittsburgh, in one of the freight yards, in a locked trailer or Cleveland. What to do? I called Ward Trucking who was associated with K & R Express and the customer service rep told me they would try to find the cabinet located somewhere but because the company went out of business nothing could be touched or containers opened until the proper authorities gave the decision to do so.

So I finally gave up on the cabinet, since I figured it was long gone at least I didn't have to pay for something that was lost in transit since it was to be delivered Freight Collect.

After several weeks had passed I got a call from Ward Trucking telling me a large box was sitting on the dock at their terminal in Altoona and would have to be picked up in a few days, and further the payment would have to be cash, no credit cards or personal checks, and lastly the price was significantly higher than what we originally agreed to. I thought this was a very unusual statement from a large eastern carrier; nevertheless, I agreed to the conditions, and drove to the site to pick up the merchandise.

Horrors

When I backed into the dock for loading, I noticed the box was not crated as I had paid for, but the cabinet was simply placed in a couple of large single cardboard boxes cobbled and taped together to fit the radio cabinet. There were no tags on the box saying "Fragile" or "This end up," so the freight company did not know the top from the bottom. I really became worried. After payment, and some discussion with the dock worker, I arrived home and unloaded the cabinet to find: lo and behold, the company (a supposed professional packaging company sanctioned by UPS) did not have protective packing around the cabinet; rather, the cabinet was shoved into the box without any protection to the grille and looked like a hurry-up and get out the door thing.

Topsy-Turvy

The cabinet had also traveled the distance upside down so that the top lid had the finish almost completely scrubbed off by the normal vibration and shifting in transit. The cabinet appeared complete though, with minimal damage through its journey to Pennsylvania other than the aforementioned top finish. I did breathe a sigh of relief.

Caveat Emptor

I am now in the process of restoring this radio and what I have learned is this: the next time I decide to purchase something large and bulky like this will be a long time coming. The total cost to date for this radio is an extraordinary amount. If I stated it here you would think it incredible for something that the original vendor might have picked up at a yard sale for a few bucks.

Sometimes when you really want something you are willing to take chances and let caution to the wind, as in this case. A lesson learned. I have bought similar items years back with few or no problems but today it seems that most everyone is in such a hurry that quality is something that time forgot.

Bibliography

All references are to *The Pittsburgh Oscillator*, 1996-2010.

Bell, Raymond M., *A Television Demonstration in 1928*, March 1999, page 7.

Brewster, Richard, *Philco Mystery Control*, June 2005, pages 8-9.

Brewster, Richard, *RCA Resurrected*, June 2008, page 8.

Depto, Ted, *Adventures in TV Land*, September 2005, pages 8-9.

Depto, Ted, *Can You Hear Me Now?,* June 2004, pages 11-13.

Depto, Ted, *Converting a Battery Tombstone to AC*, March 2002, pages 5-6.

Depto, Ted and Mike Simpson, *The Midwest Story,*
    Part I: June 2001, pages 5, 6, 8, 9
    Part II: September 2001, pages 7, 8, 13
    Part III: December 2001, pages 7-10.

Depto, Ted, *The Philco Model 41-608H*, September 1999, pages 7-8.

Flaherty, Rege, *My Philco 90*, March 2007, page 12.

Gaetano, Lou, *Should I Plug in the Old Radio I Found at the Flea Market?*, March 2000, page 5.

Garland, Jim and Don Merz, *The Coming Deluge: A Brief Debate*, December 1999, pages 5-6.

Haught, John W., *On the Road With John and Sara*, June 1997, page 10.

Hyman, Paul R., *A Hopeless Case*, December 2007, page 14.

Hyman, Paul R., *Restoration Corner*, September 1998, pages 7-8.

Hyman, Paul R., *Small Receiving Antennas*, March 2008, pages 12-14

Hyman, Paul R., *Zenith Transoceanic 1L6's*, June 1998, page 7.

Ivarson, David C., *The Majestic That Was*, December 1998, page 7.

Lindenbach, Walter, *CD as Hallucinogen*, June 2006, pages 10-12.

Male, Michael, *A Catalin Saturday in Clarksville*, March 2002, pages 8-9.

McGuigan, Ed, *When Ed Speaks–My First Radio*, June 2006, page 7.

Merz, Don and Jim Garland, *The Coming Deluge: A Brief Debate*, December 1999, pages 5-6.

Mizikar, Tom, *Tom Mizikar Writes About His FADA TV*, December 2008, page 20.

Patrick, Joe, *Breting 12 Receiver*, September 2010, pages 6-9.

Patrick, Joe, *'Hands-on' Troubleshooting*, June 2010, pages 5-8.

Paul, Floyd A., *The Kemper Radio Corporation*, June 2005, pages 12-16.

Piesik, Dan, *Golden Age Radio*, June 2007, page 13.

Simpson, Mike and Ted Depto, *The Midwest Story,*
    Part I: June 2001, pages 5, 6, 8, 9
    Part II: September 2001, pages 7, 8, 13
    Part III: December 2001, pages 7-10.

Togyer, Jason, *Radio Days*, March 1999, pages 8-9.

Wells, Chris, *Plaskon Radios*, June 2009, page 6.

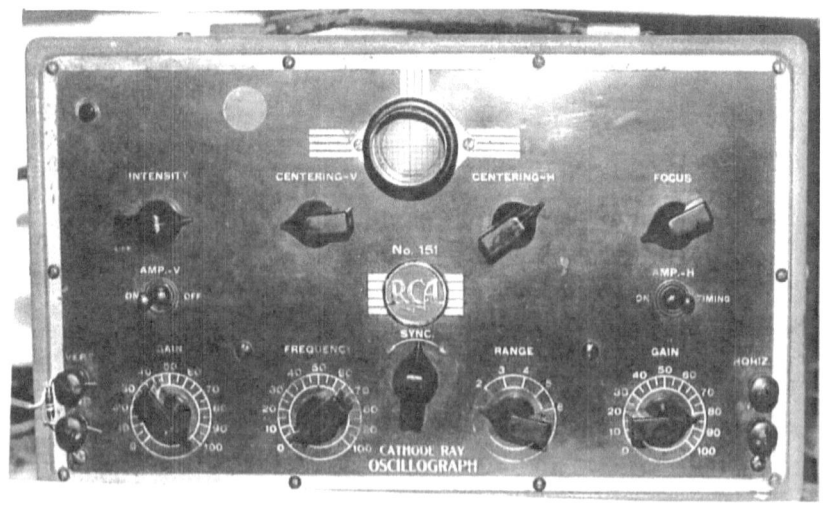

**Tom Dixon's restored RCA one-inch oscilloscope
drew a lot of interest at the June 2007 PARS meet.**

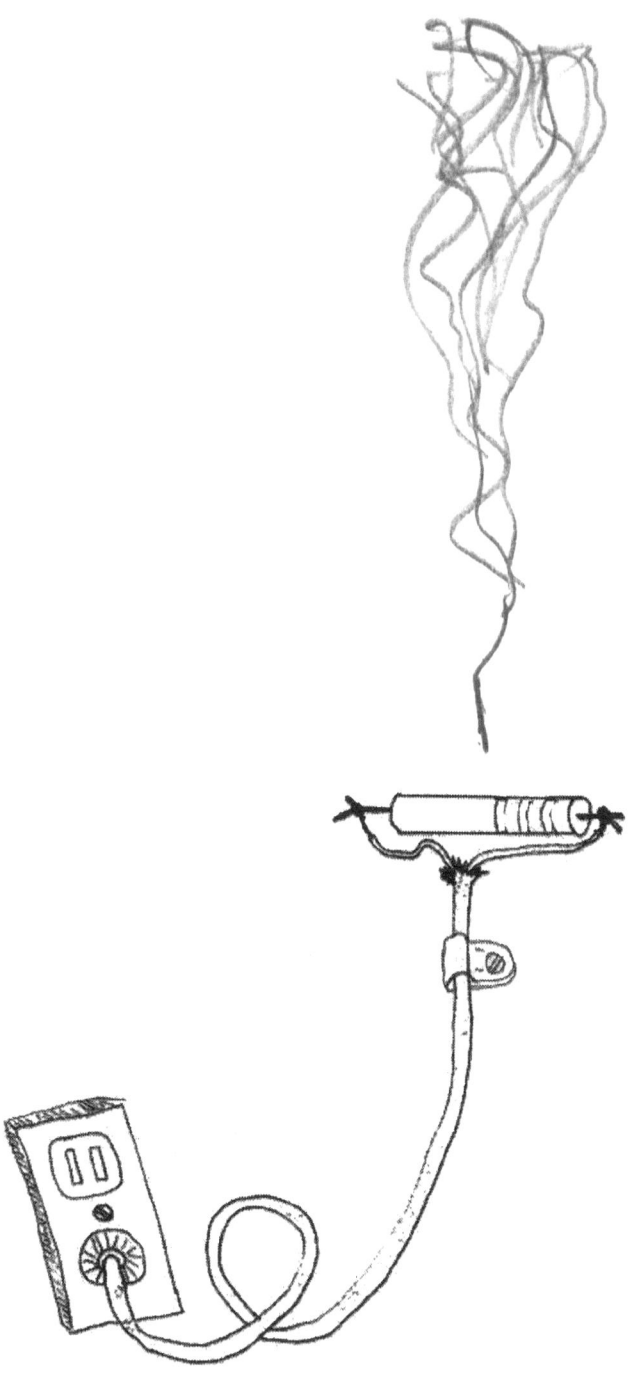

I=E/R   R=E/I   E=IR   P=IE